Photomicrographs of invertebrates

A. C. Shaw
Head of the Biology Department, The Skinners' School, Tunbridge Wells

S. K. Lazell, A.R.P.S.
Medical Photographer, Tunbridge Wells Hospital Group

G. N. Foster
Lecturer in the Zoology Department,
The West of Scotland Agricultural College, Ayr

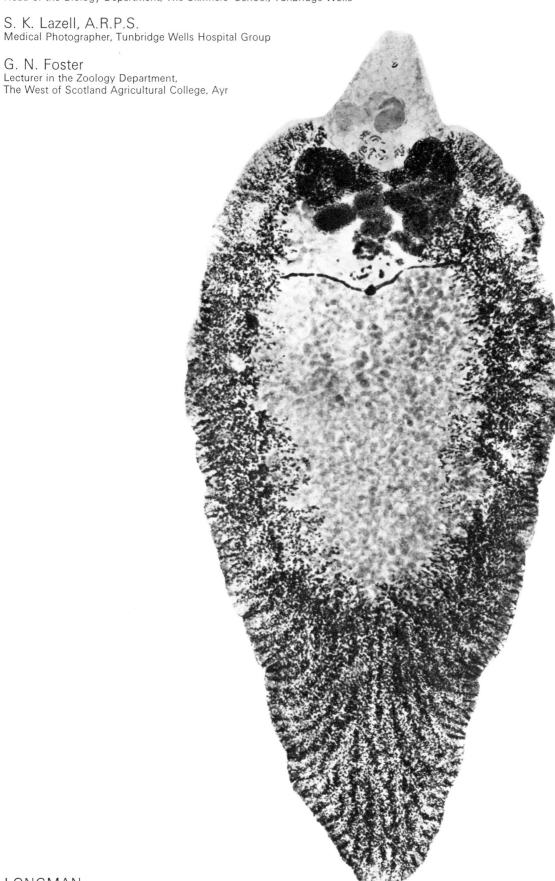

LONGMAN

Longman
1724-1974

Longman Group Limited London
Associated companies, branches and representatives throughout the world

First published 1974
ISBN 0 582 32279 0

Printed in Great Britain by
Fletcher & Son Ltd, Norwich

Contents

Preface

The success of *Photomicrographs of the Flowering Plant* and *Photomicrographs of the Non-flowering Plant* has encouraged us to produce a similar work on the structure and reproduction of invertebrates. The aim of this series is to present to students of zoology, botany and biology at Advanced Level a set of photomicrographs side by side with labelled diagrams so that they may interpret what they see under the light microscope.

Wherever possible, we have chosen slides of the standard normally available to schools. The types of nearly all Advanced Level syllabuses are represented, together with material used in courses at technical colleges.

We have provided many sections of the earthworm and the cockroach because we believe that detailed analysis of sections of small animals is an excellent way of familiarising students with their anatomy. Slides of cockroach sections are not yet readily available from supply agencies ; ours were cut from a recently-moulted cockroach nymph by members of the pathology department of our local hospital.

The photographs were obtained with a Beck London 47 microscope and eyepiece camera using Ilford Micro-Neg Pan. The drawings were prepared by inking over a faint print and then bleaching out the photograph.

Tunbridge Wells, 1972

A. C. Shaw
S. K. Lazell
G. N. Foster

Acknowledgements

Our slides were purchased mainly from Gerrard & Haig Ltd, Harris Biological Supplies Ltd and GBI (Labs) Ltd. We wish to thank Mr T. A. M. Gerrard and his colleagues who prepared a special slide for us, and Mr Eric Tatchell who provided slides of *Plasmodium*.

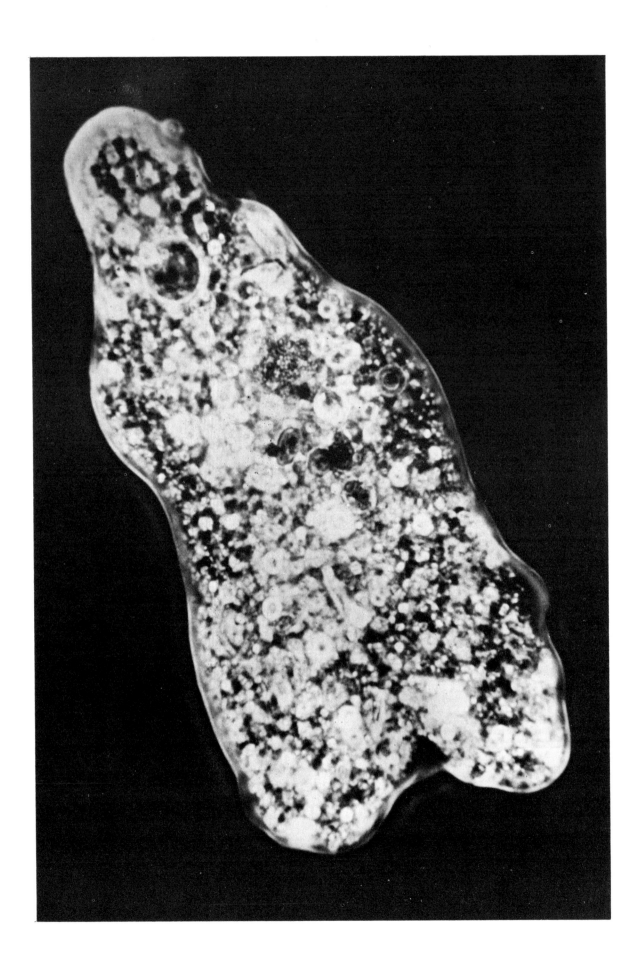

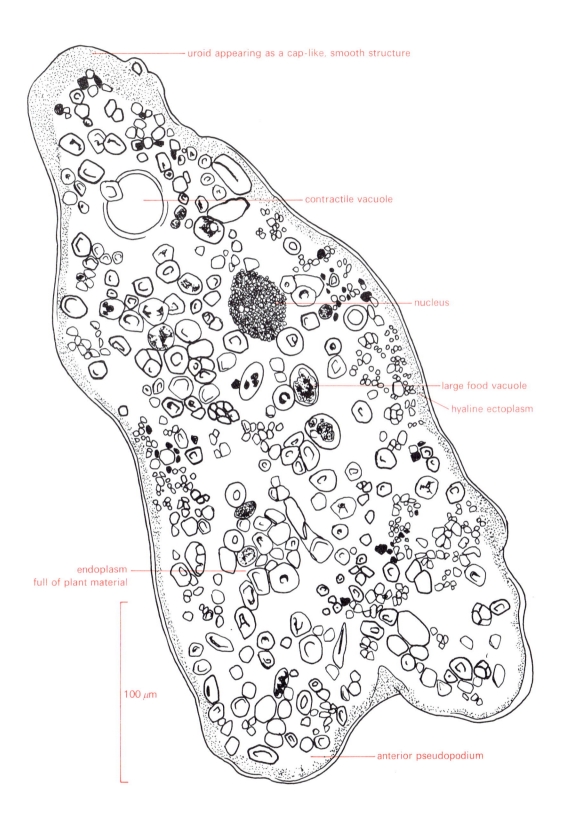

Fig. 1. *Amoeba* full of undigested plant material – phase contrast
photomicrograph by Mr M. I. Walker
© Harris Biological Supplies Ltd.

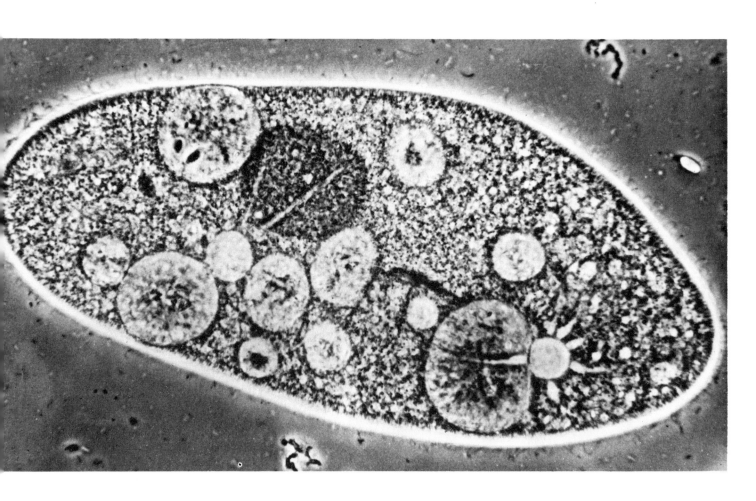

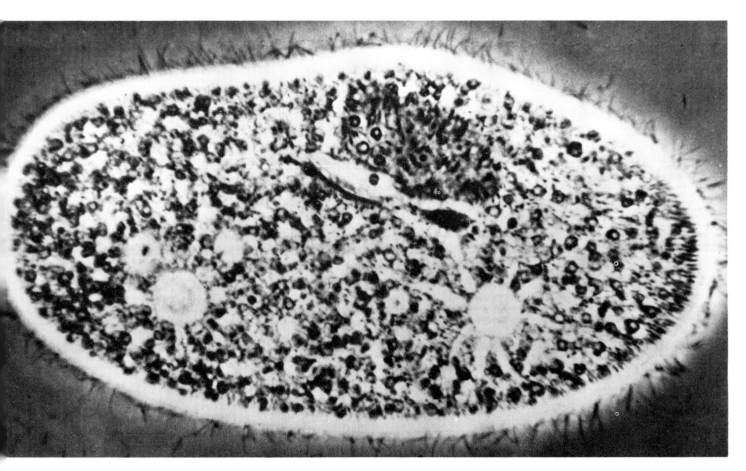

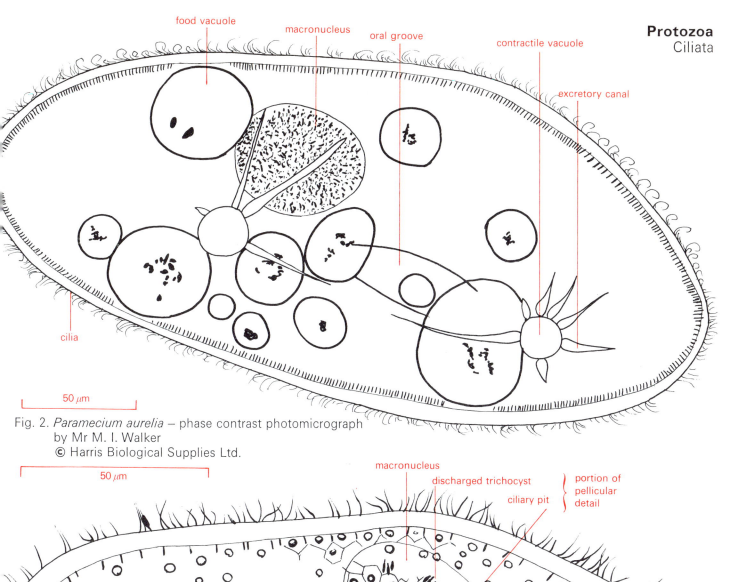

food vacuole · macronucleus · oral groove · contractile vacuole · excretory canal · cilia · 50 μm

Fig. 2. *Paramecium aurelia* – phase contrast photomicrograph
by Mr M. I. Walker
© Harris Biological Supplies Ltd.

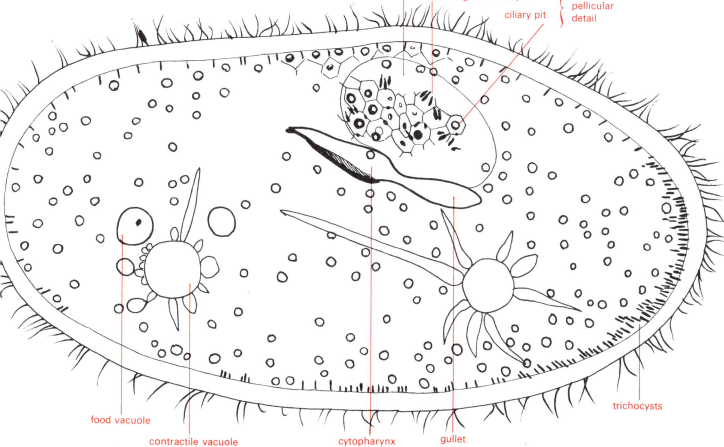

50 μm · macronucleus · discharged trichocyst · ciliary pit · portion of pellicular detail · trichocysts · food vacuole · contractile vacuole · cytopharynx · gullet

Fig. 3. *Paramecium bursaria* – phase contrast photomicrograph
by Mr M. I. Walker
© Harris Biological Supplies Ltd.

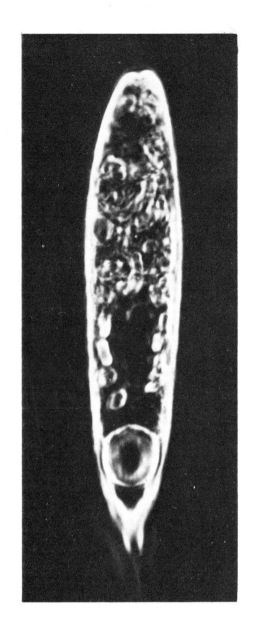

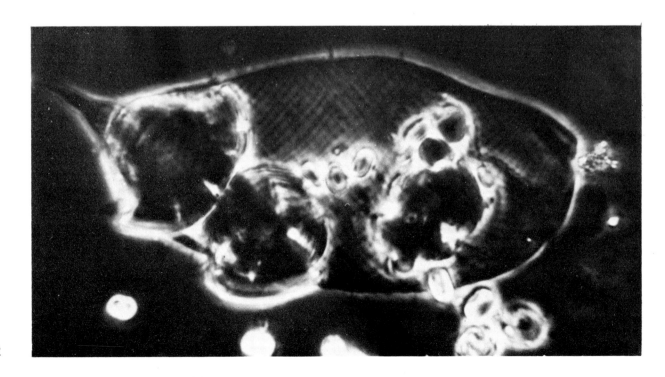

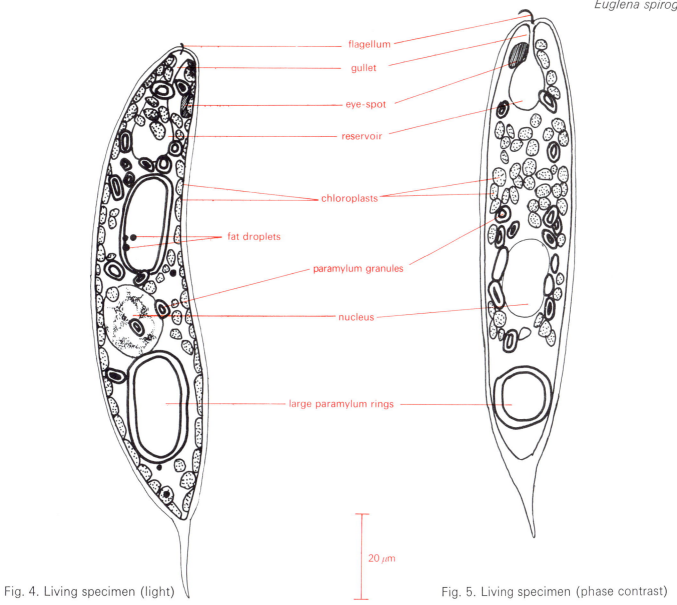

flagellum

gullet

eye-spot

reservoir

chloroplasts

fat droplets

paramylum granules

nucleus

large paramylum rings

20 μm

Fig. 4. Living specimen (light)

Fig. 5. Living specimen (phase contrast)

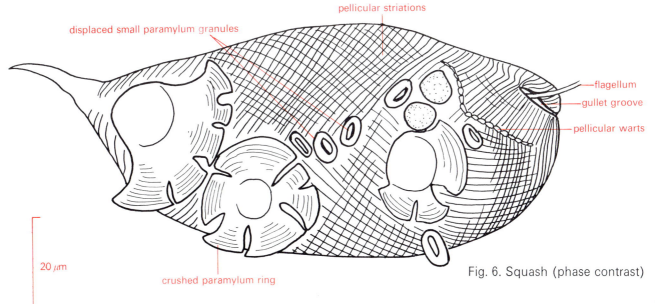

pellicular striations

displaced small paramylum granules

flagellum

gullet groove

pellicular warts

20 μm

crushed paramylum ring

Fig. 6. Squash (phase contrast)

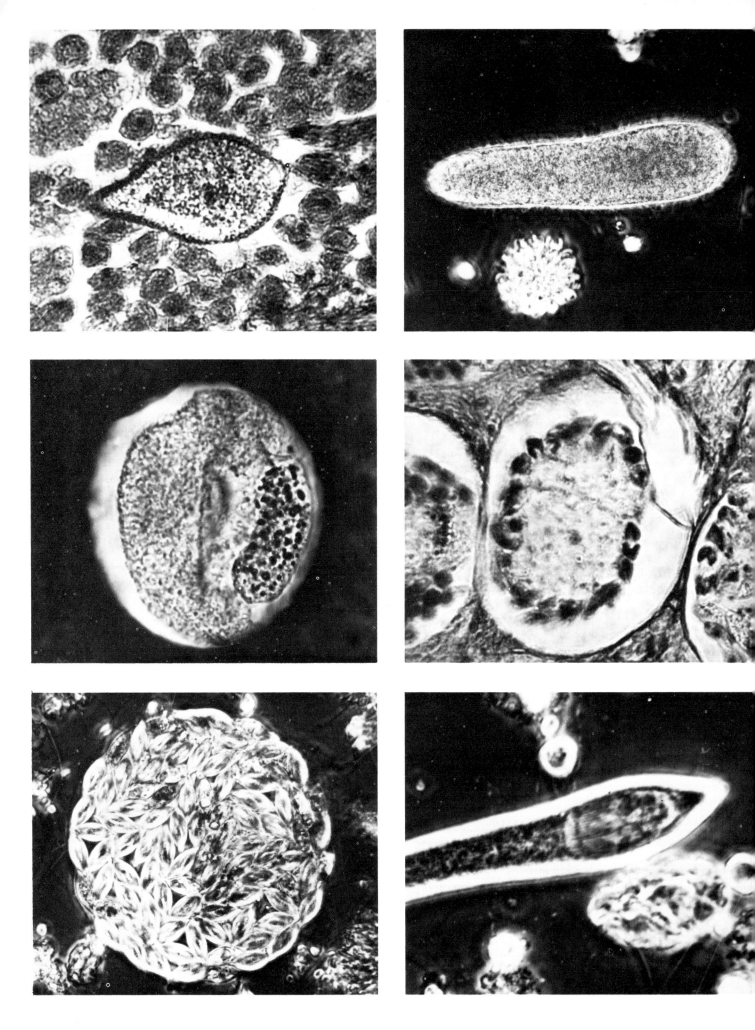

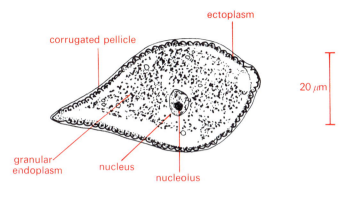

Fig. 7. *Monocystis* – V.S. trophozoite in section of earthworm seminal vesicles

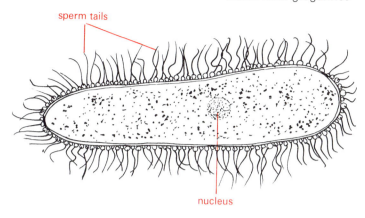

Fig. 8. *Monocystis* – trophozoite – phase contrast

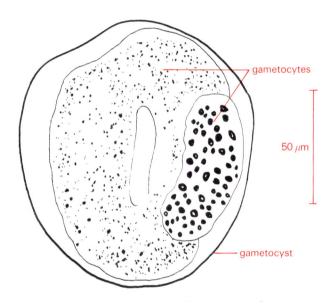

Fig. 9. *Monocystis* – gametocyst (gametocytes)

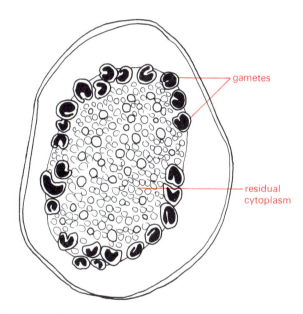

Fig. 10. *Monocystis* – gametocyst (gametes)

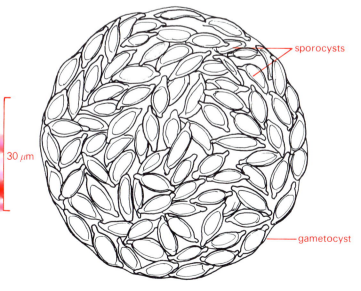

Fig. 11. *Monocystis* – gametocyst (sporocysts) – phase contrast

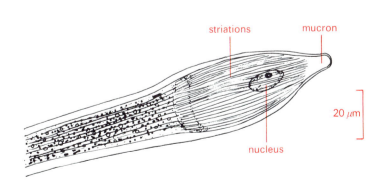

Fig. 12. *Rhyncocystis* – from earthworm (phase contrast)

15

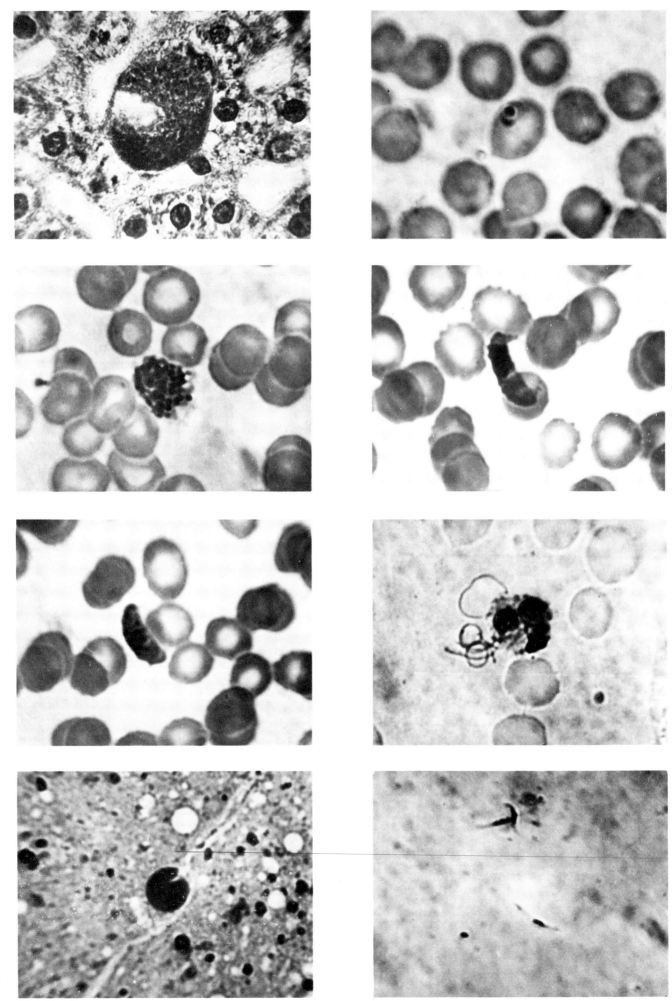

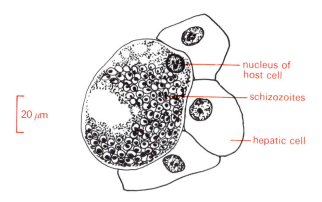

20 μm

nucleus of
host cell

schizozoites

hepatic cell

Fig. 13. *Plasmodium vivax* – schizogony in human
liver cell (pre-erythrocytic stage)

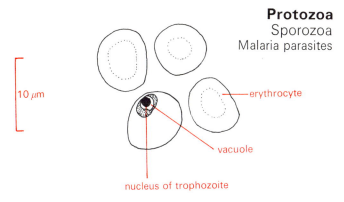

10 μm

erythrocyte

vacuole

nucleus of trophozoite

Fig. 14. *Plasmodium vivax* – ring form of
trophozoite in human blood smear

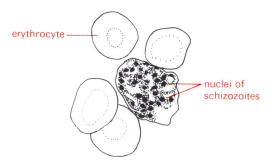

erythrocyte

nuclei of
schizozoites

Fig. 15. *Plasmodium vivax* – schizogony in human
blood smear

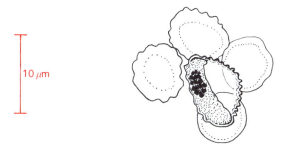

10 μm

Fig. 16. *Plasmodium falciparum* – microgametocyte
in human blood smear

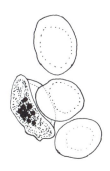

Fig. 17. *Plasmodium falciparum* – microgametocyte
in human blood smear

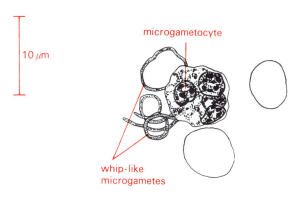

10 μm

microgametocyte

whip-like
microgametes

Fig. 18. *Plasmodium vivax* – microgamete
production in mosquito gut smear

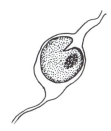

Fig. 19. *Plasmodium vivax* – ookinete in mosquito
gut wall

10 μm

Fig. 20. *Plasmodium vivax* – sporozoites from
mosquito salivary gland

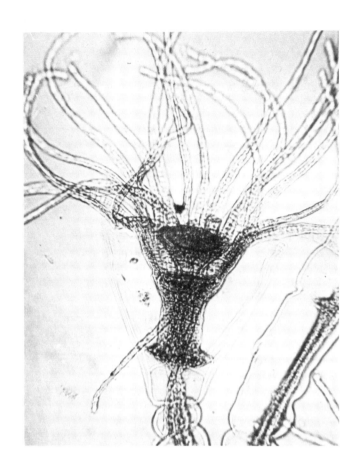

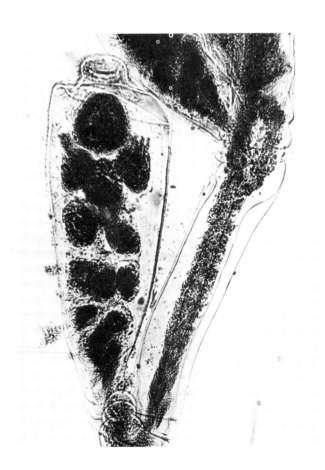

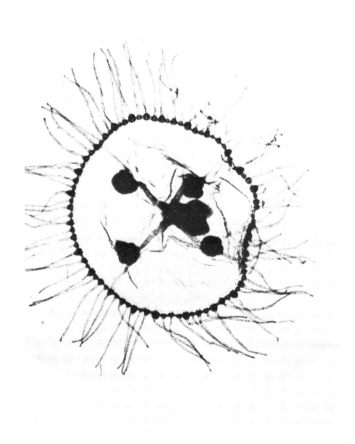

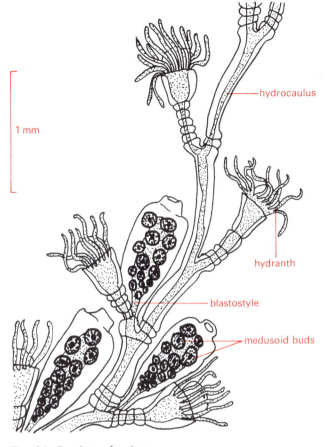

Fig. 21. Portion of colony

hydrocaulus

hydranth

blastostyle

medusoid buds

1 mm

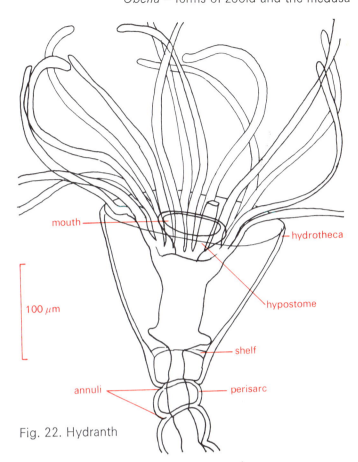

Fig. 22. Hydranth

mouth

hydrotheca

hypostome

shelf

annuli

perisarc

100 μm

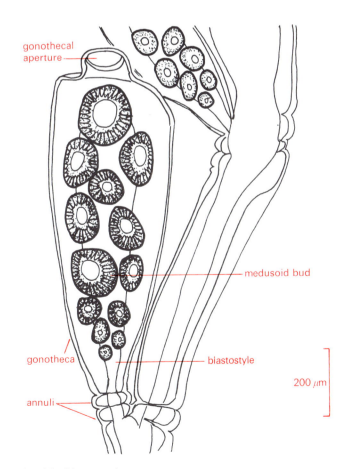

Fig. 23. Blastostyle

gonothecal aperture

medusoid bud

gonotheca

blastostyle

annuli

200 μm

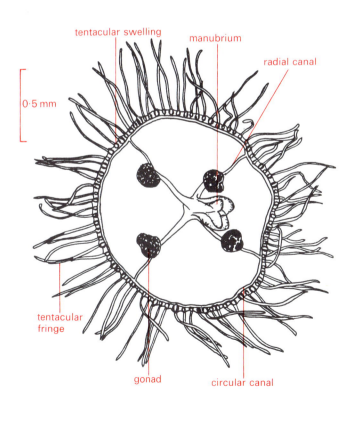

Fig. 24. Medusa

tentacular swelling

manubrium

radial canal

tentacular fringe

gonad

circular canal

0·5 mm

19

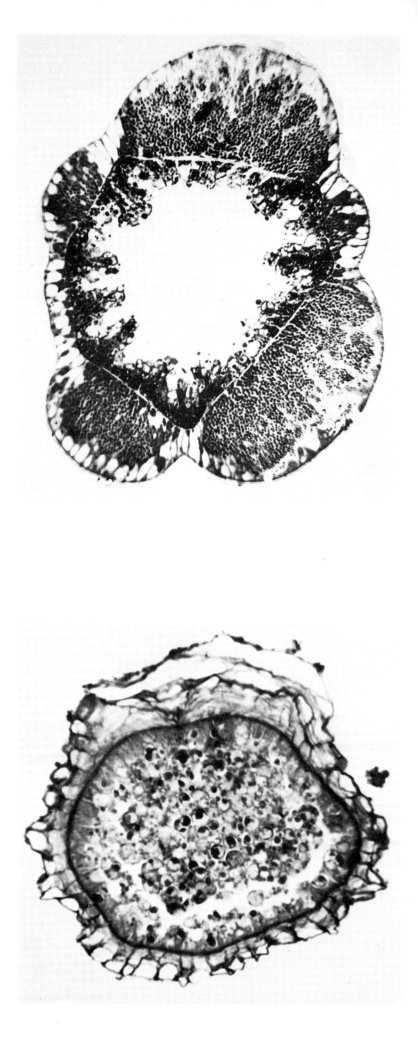

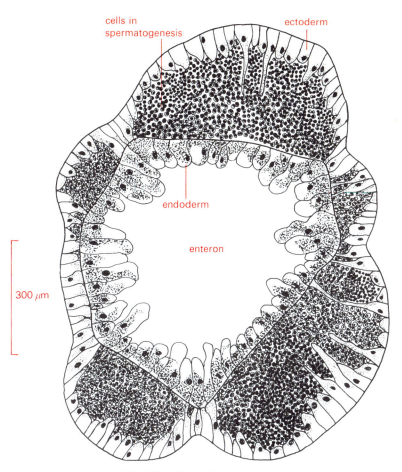

Fig. 26. T.S. region of testes

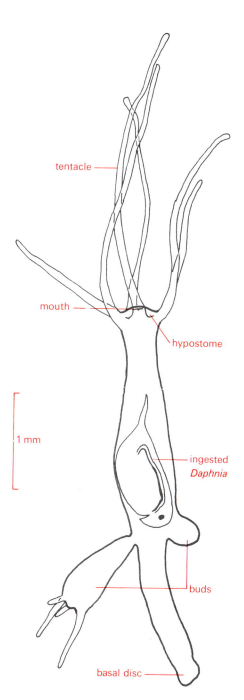

Fig. 25. Entire bud

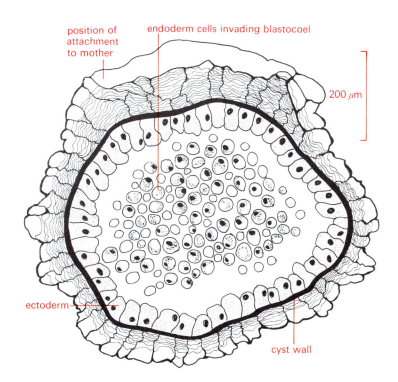

Fig. 27. Section of developing gastrula
within cyst

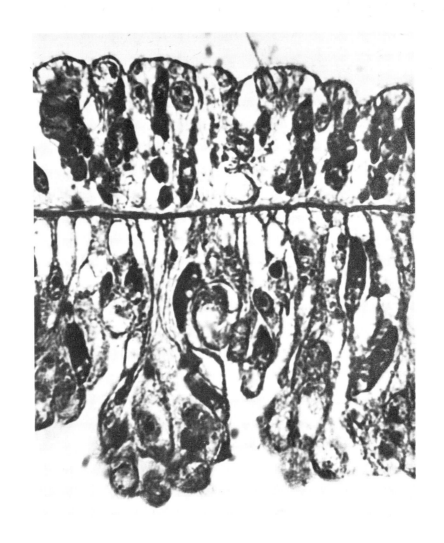

Coelenterata
Hydrozoa
Hydra – body wall (Geimsa's eosin
azure stain for I cells)

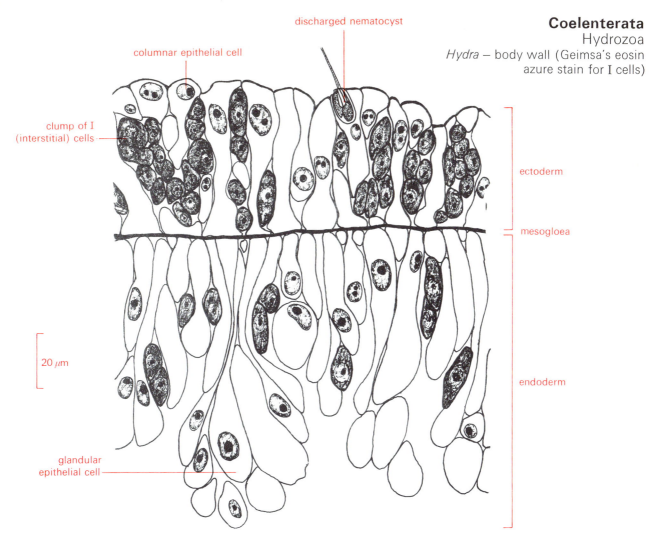

discharged nematocyst

columnar epithelial cell

clump of I
(interstitial) cells

ectoderm

mesogloea

20 μm

endoderm

glandular
epithelial cell

Fig. 28. T.S. body wall

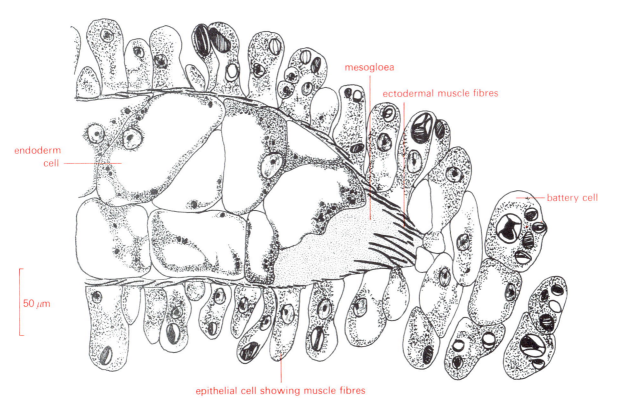

mesogloea

ectodermal muscle fibres

endoderm
cell

battery cell

50 μm

epithelial cell showing muscle fibres

Fig. 29. Grazing section of tentacle

23

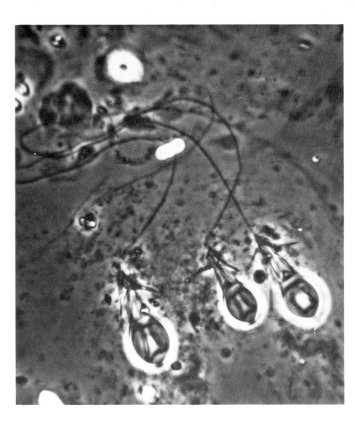
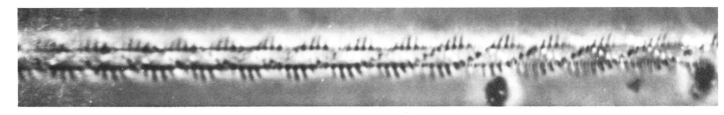
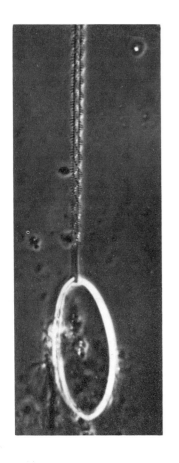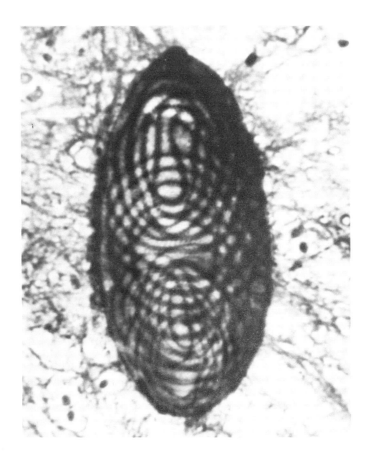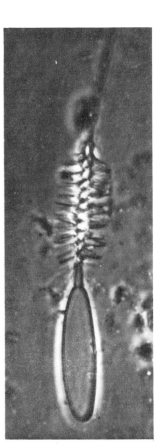

Coelenterata
Hydrozoa ; Anthozoa
Nematocysts : the giant nematocysts of the cup coral, *Corynactis*,
provide excellent material for classwork

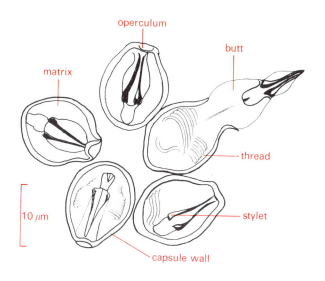

Fig. 30. *Hydra* – undischarged and partly discharged
stenoteles (penetrants) – phase contrast

Fig. 31. *Hydra* – fully discharged stenoteles –
phase contrast

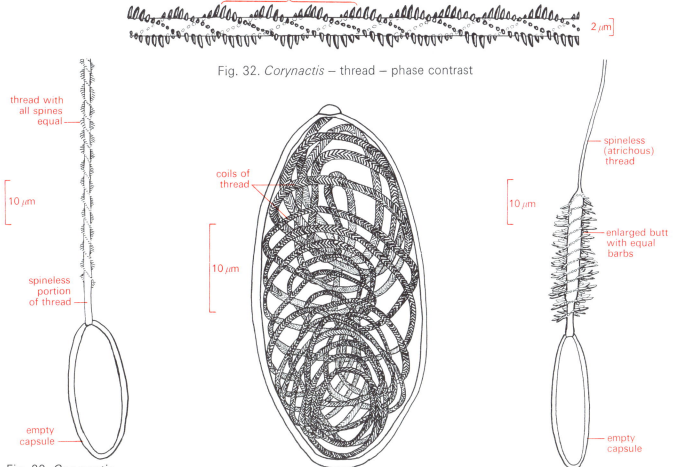

Fig. 32. *Corynactis* – thread – phase contrast

Fig. 33. *Corynactis* –
discharged
holotrichous isorhiza–
phase contrast

Fig. 34. *Corynactis* –
undischarged isorhiza –
light

Fig. 35. *Corynactis* – discharged microbasic
(bottle brush) – phase contrast
mastigophore

25

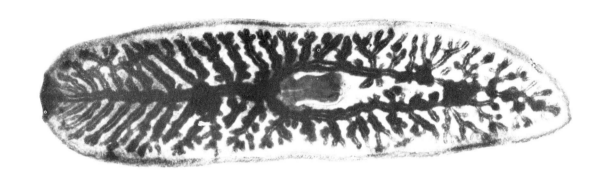

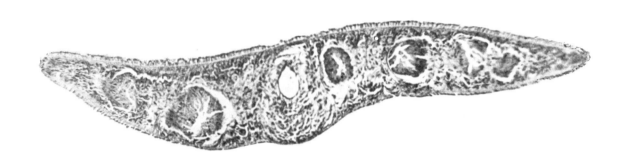

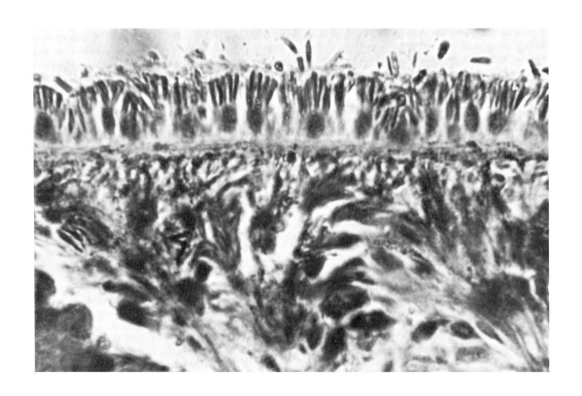

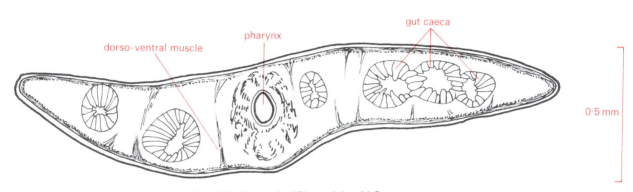

three main branches of gut

gut caeca

minute horn

eyes

1 mm

eversible pharynx

genital atrium

Fig. 36. *Dendrocoelum* – entire

gut caeca

dorso-ventral muscle

pharynx

0·5 mm

Fig. 37. *Dugesia* (*Planaria*) – V.S.

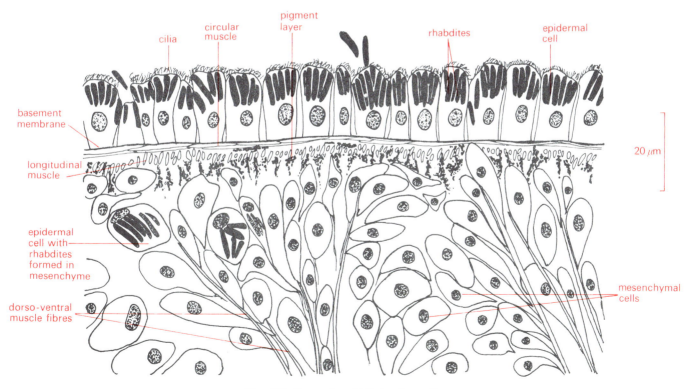

cilia

circular muscle

pigment layer

rhabdites

epidermal cell

basement membrane

20 μm

longitudinal muscle

epidermal cell with rhabdites formed in mesenchyme

mesenchymal cells

dorso-ventral muscle fibres

Fig. 38. *Dugesia* – V.S. body wall

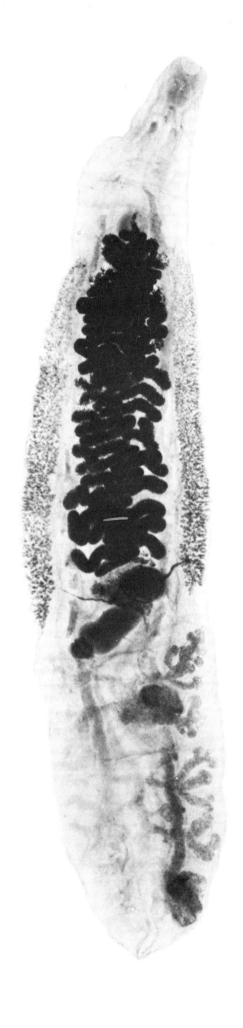

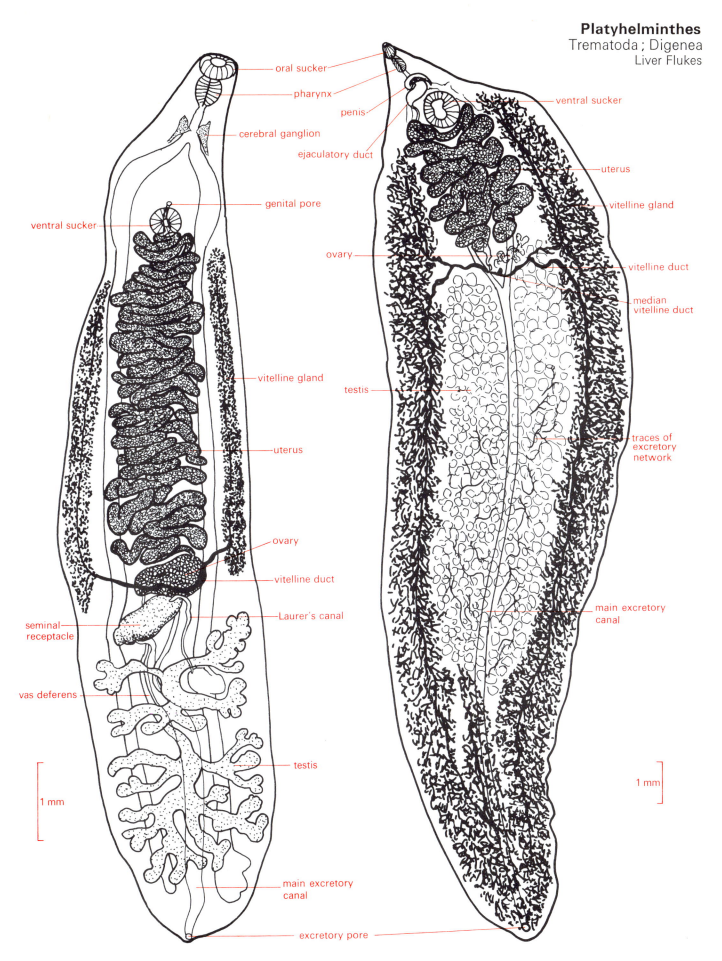

oral sucker
pharynx
penis
cerebral ganglion
ejaculatory duct
genital pore
ventral sucker
ventral sucker
uterus
vitelline gland
ovary
vitelline duct
median vitelline duct
vitelline gland
testis
uterus
ovary
vitelline duct
Laurer's canal
traces of excretory network
seminal receptacle
vas deferens
testis
main excretory canal
main excretory canal
excretory pore

1 mm

1 mm

Fig. 39. *Clonorchis sinensis* (Chinese liver fluke) – entire   Fig. 40. *Fasciola hepatica* (Sheep liver fluke) – entire

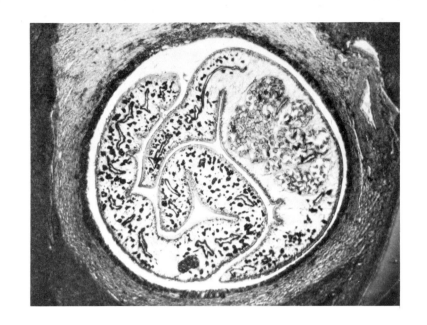

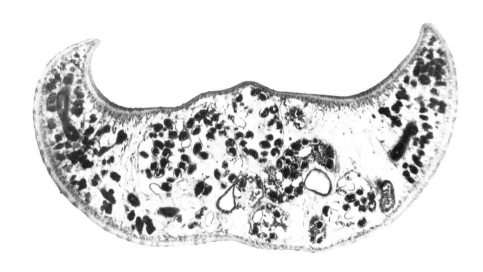

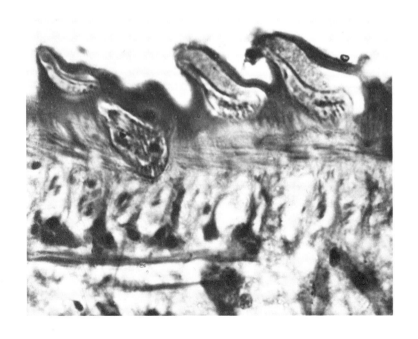

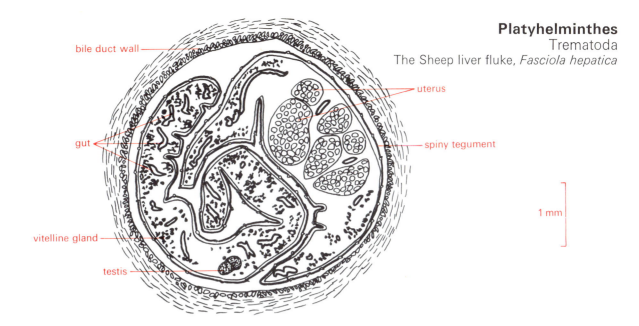

bile duct wall

uterus

gut

spiny tegument

vitelline gland

testis

1 mm

Fig. 41. Section of liver flukes in bile duct of sheep

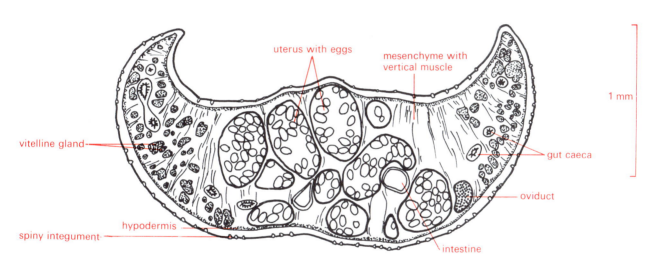

uterus with eggs

mesenchyme with vertical muscle

vitelline gland

gut caeca

oviduct

hypodermis

spiny integument

intestine

1 mm

Fig. 42. V.S. body

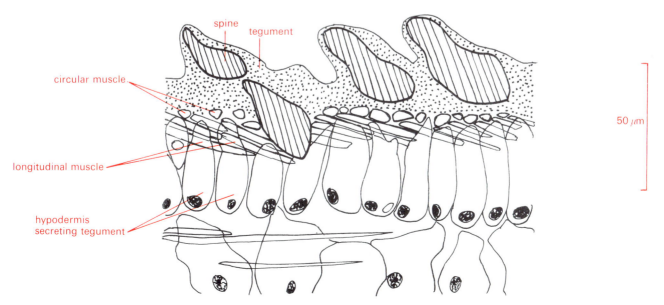

spine

tegument

circular muscle

longitudinal muscle

hypodermis secreting tegument

50 μm

Fig. 43. L.S. tegument

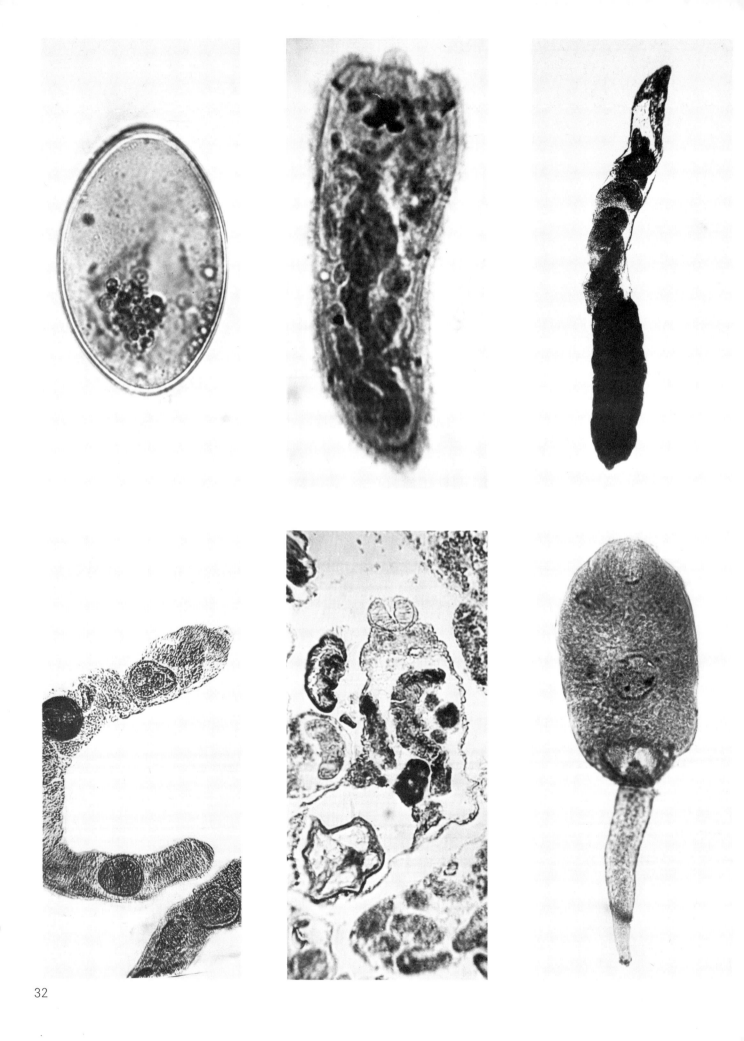

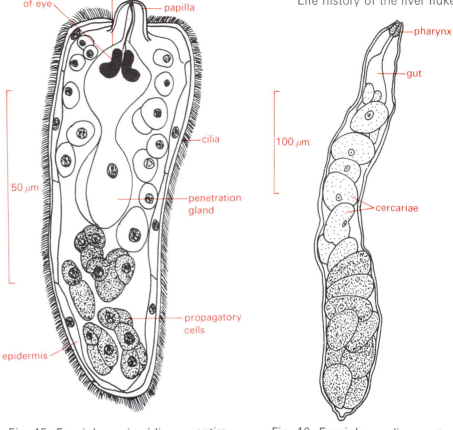

Fig. 44. *Fasciola* – egg – entire

Fig. 45. *Fasciola* – miracidium – entire

Fig. 46. *Fasciola* – redia – entire

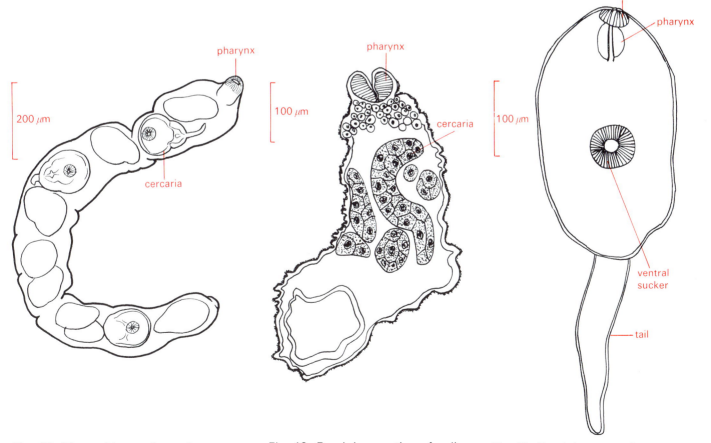

Fig. 47. *Clonorchis* – redia, entire

Fig. 48. *Fasciola* – section of redia in digestive gland of snail

Fig. 49. *Fasciola* – cercaria

33

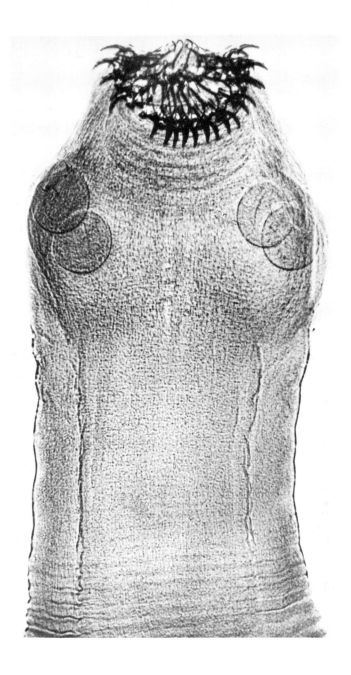

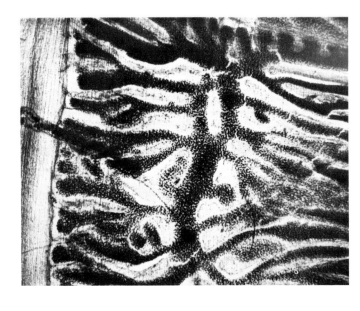

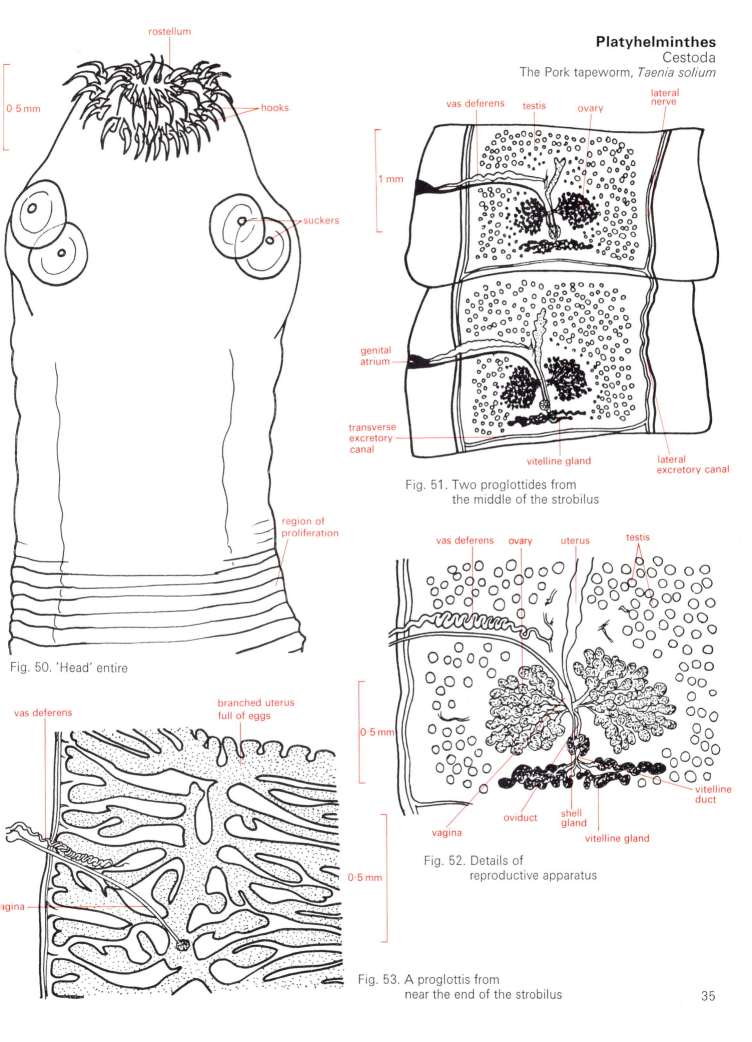

rostellum

0·5 mm

hooks

suckers

region of proliferation

Fig. 50. 'Head' entire

Platyhelminthes
Cestoda
The Pork tapeworm, *Taenia solium*

vas deferens   testis   ovary   lateral nerve

1 mm

genital atrium

transverse excretory canal

vitelline gland

lateral excretory canal

Fig. 51. Two proglottides from the middle of the strobilus

vas deferens   ovary   uterus   testis

0·5 mm

vagina   oviduct   shell gland   vitelline gland   vitelline duct

Fig. 52. Details of reproductive apparatus

vas deferens   branched uterus full of eggs

0·5 mm

0·5 mm

vagina

Fig. 53. A proglottis from near the end of the strobilus

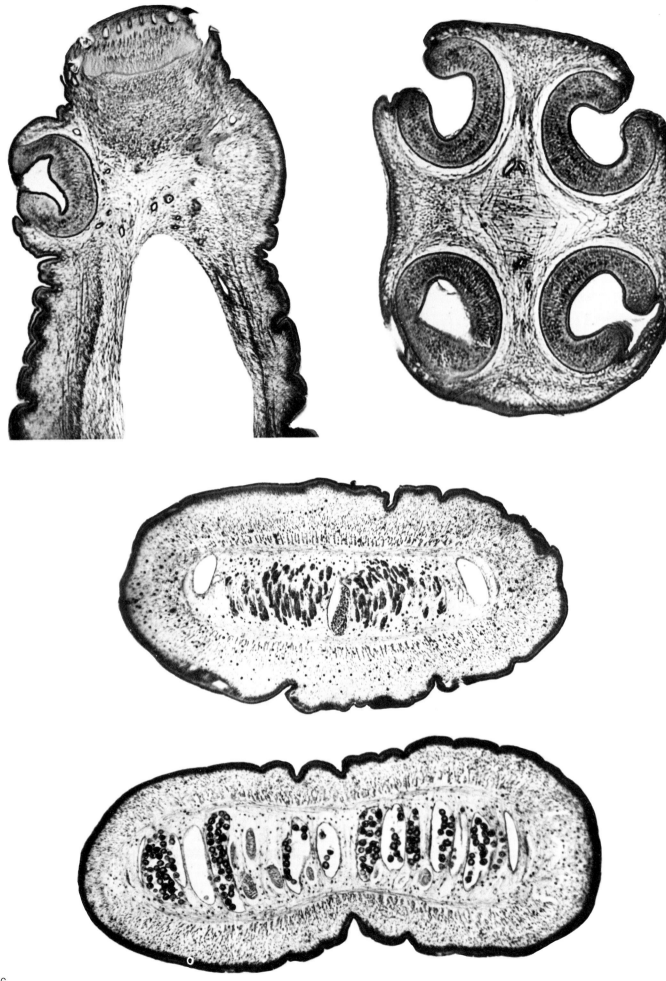

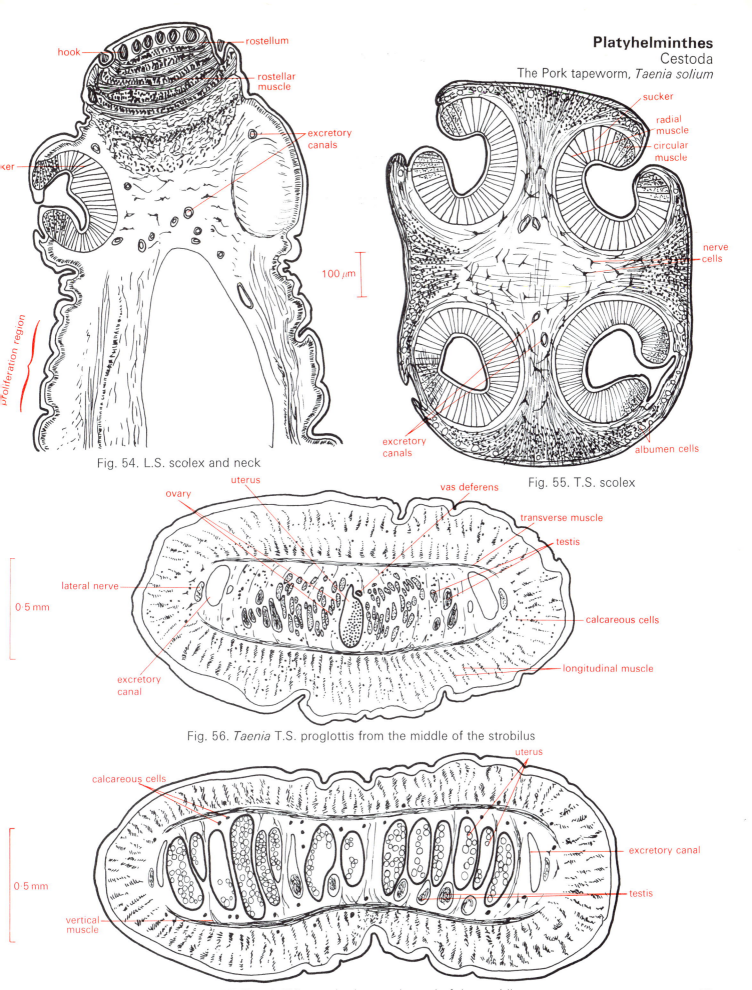

hook

rostellum

rostellar muscle

excretory canals

ker

proliferation region

100 μm

Fig. 54. L.S. scolex and neck

sucker

radial muscle

circular muscle

nerve cells

excretory canals

albumen cells

Fig. 55. T.S. scolex

ovary

uterus

vas deferens

transverse muscle

testis

lateral nerve

0·5 mm

calcareous cells

excretory canal

longitudinal muscle

Fig. 56. *Taenia* T.S. proglottis from the middle of the strobilus

calcareous cells

uterus

excretory canal

0·5 mm

testis

vertical muscle

Fig. 57. *Taenia* T.S. proglottis near the end of the strobilus

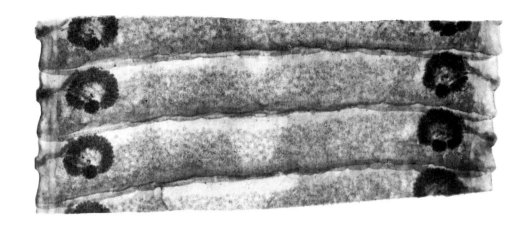

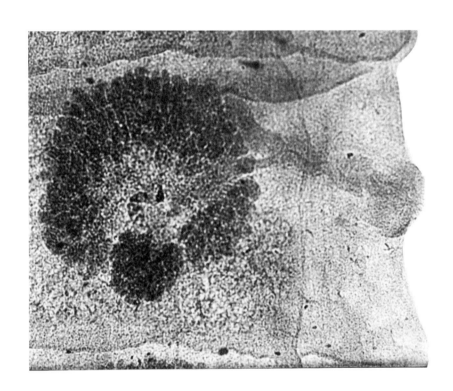

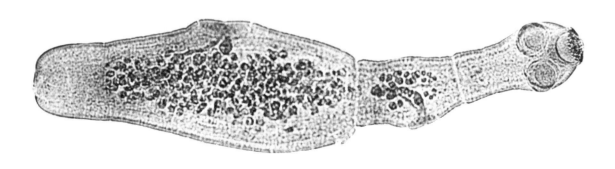

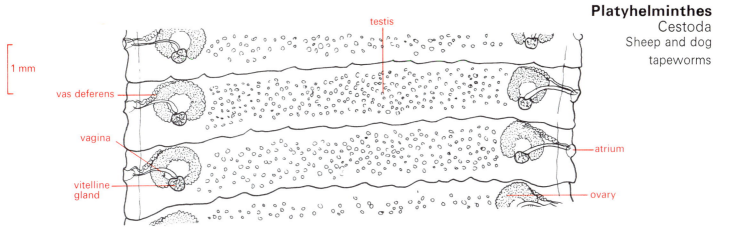

Fig. 58. *Moniezia* (sheep tapeworm) – proglottides from the middle of the strobilus

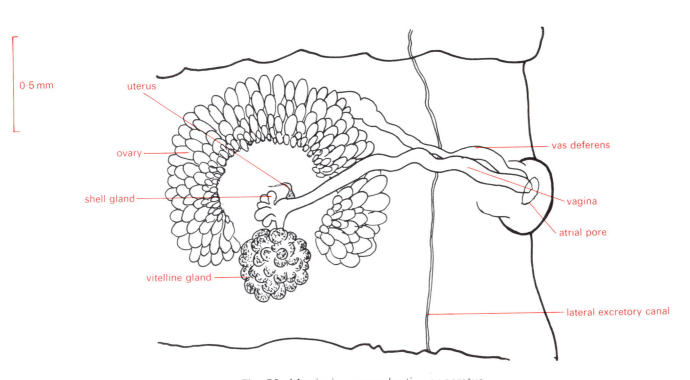

Fig. 59. *Moniezia* – reproductive apparatus

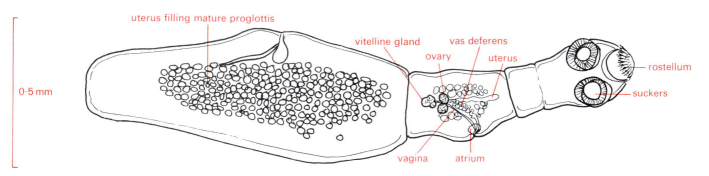

Fig. 60. *Echinococcus* (dog tapeworm) – entire worm

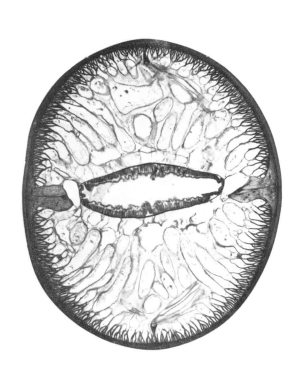

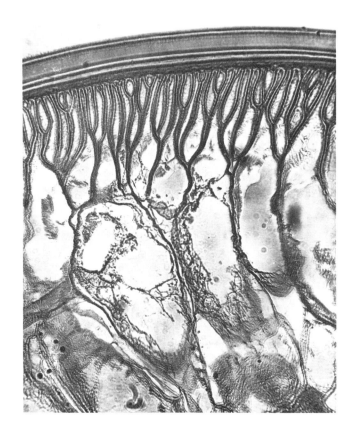

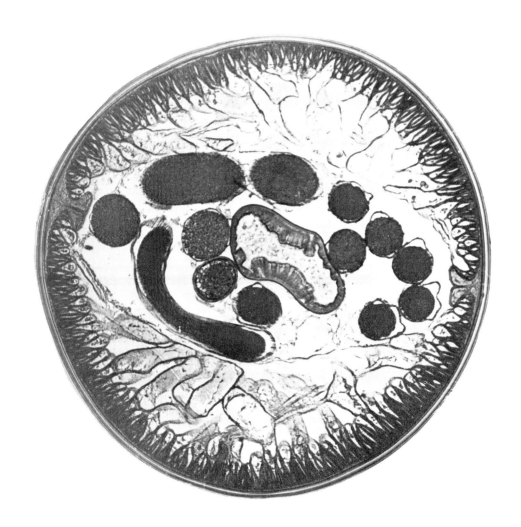

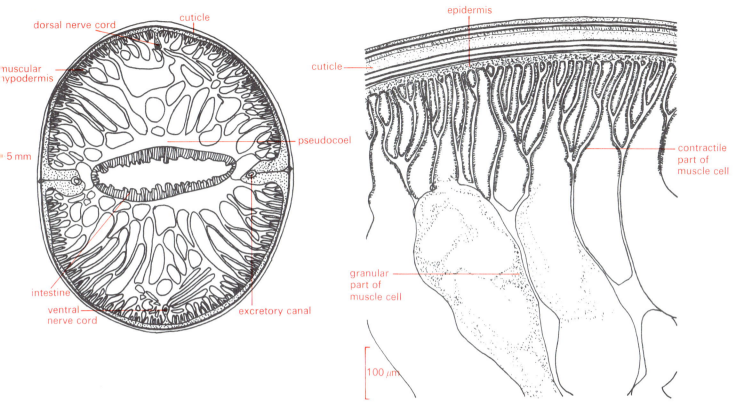

Fig. 61. T.S. anterior region

Fig. 62. T.S. body wall

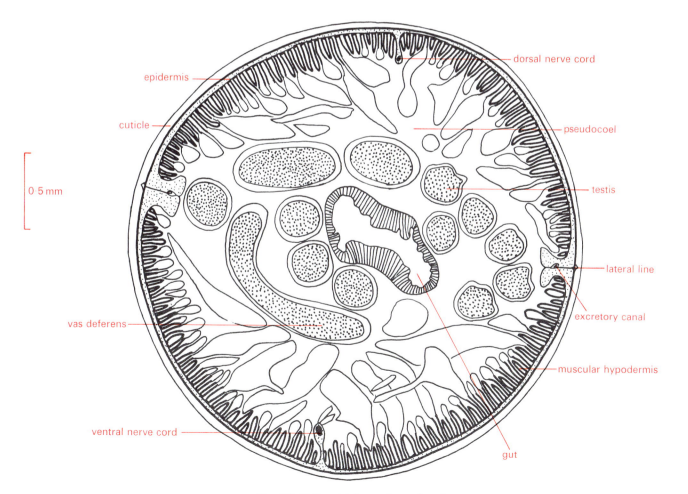

Fig. 63. T.S. middle region of male

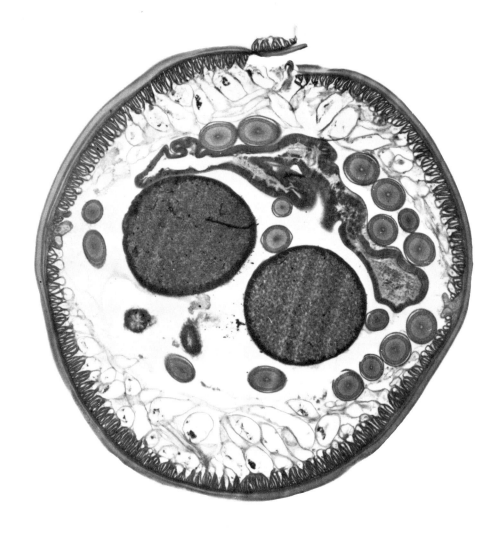

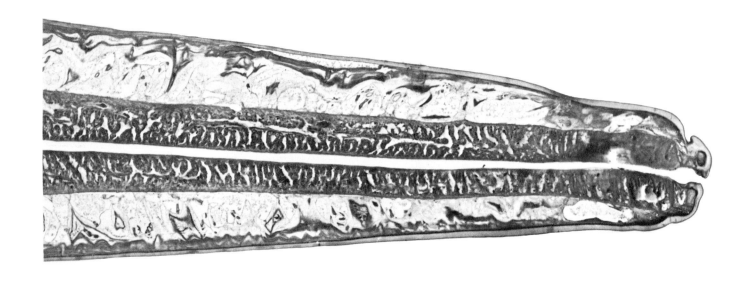

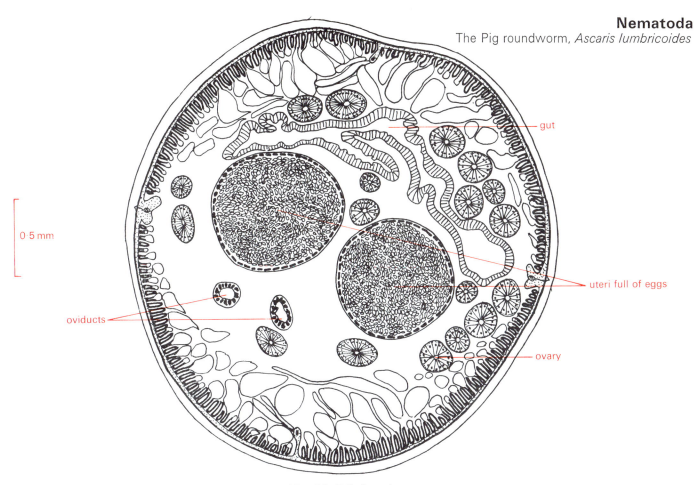

0·5 mm

gut

uteri full of eggs

oviducts

ovary

Fig. 64. T.S. female

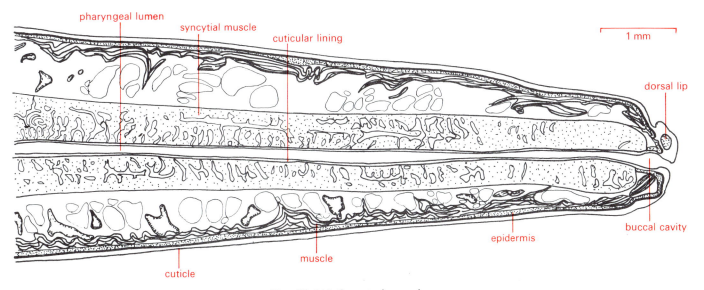

pharyngeal lumen   syncytial muscle   cuticular lining

1 mm

dorsal lip

buccal cavity

epidermis

cuticle   muscle

Fig. 65. V.L.S. anterior region

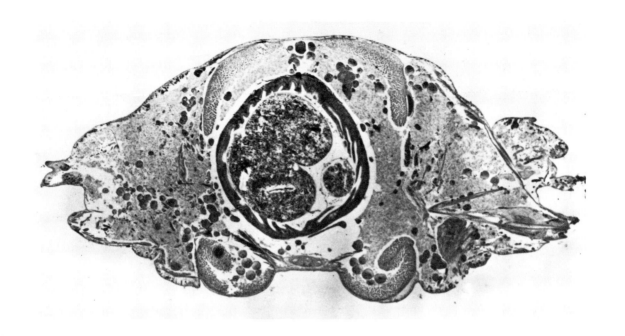

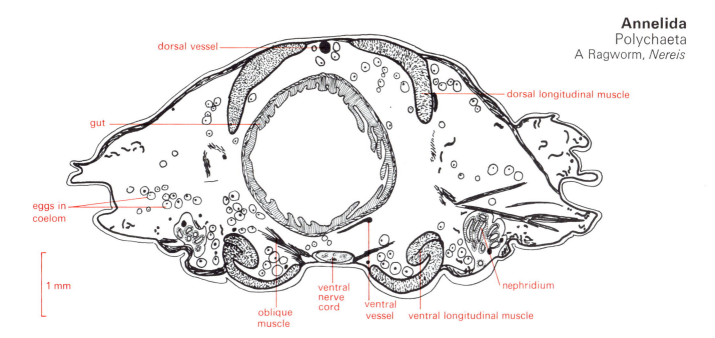

dorsal vessel

dorsal longitudinal muscle

gut

eggs in
coelom

1 mm

oblique
muscle

ventral
nerve
cord

ventral
vessel

ventral longitudinal muscle

nephridium

Fig. 66. T.S. segment

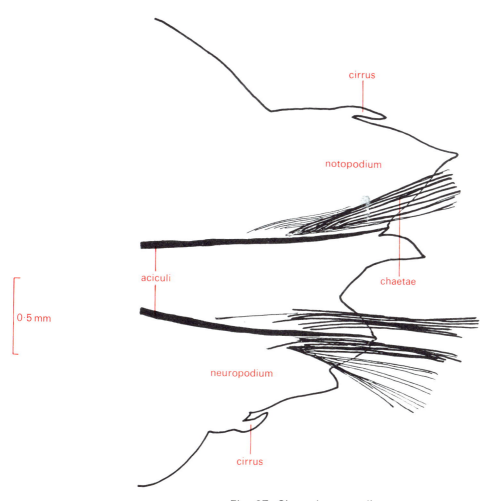

cirrus

notopodium

aciculi

chaetae

0·5 mm

neuropodium

cirrus

Fig. 67. Cleared parapodium

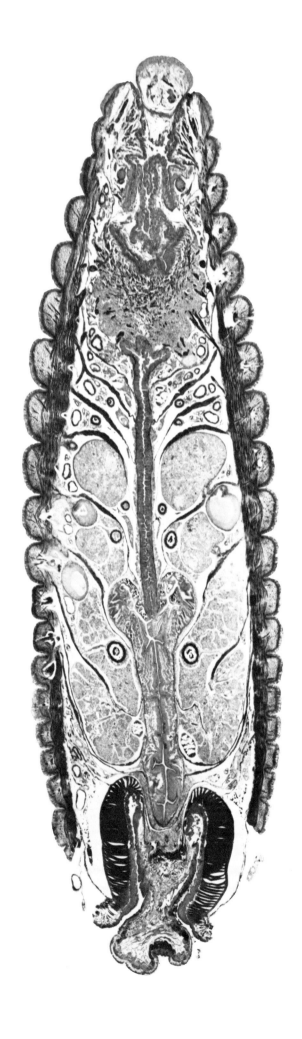

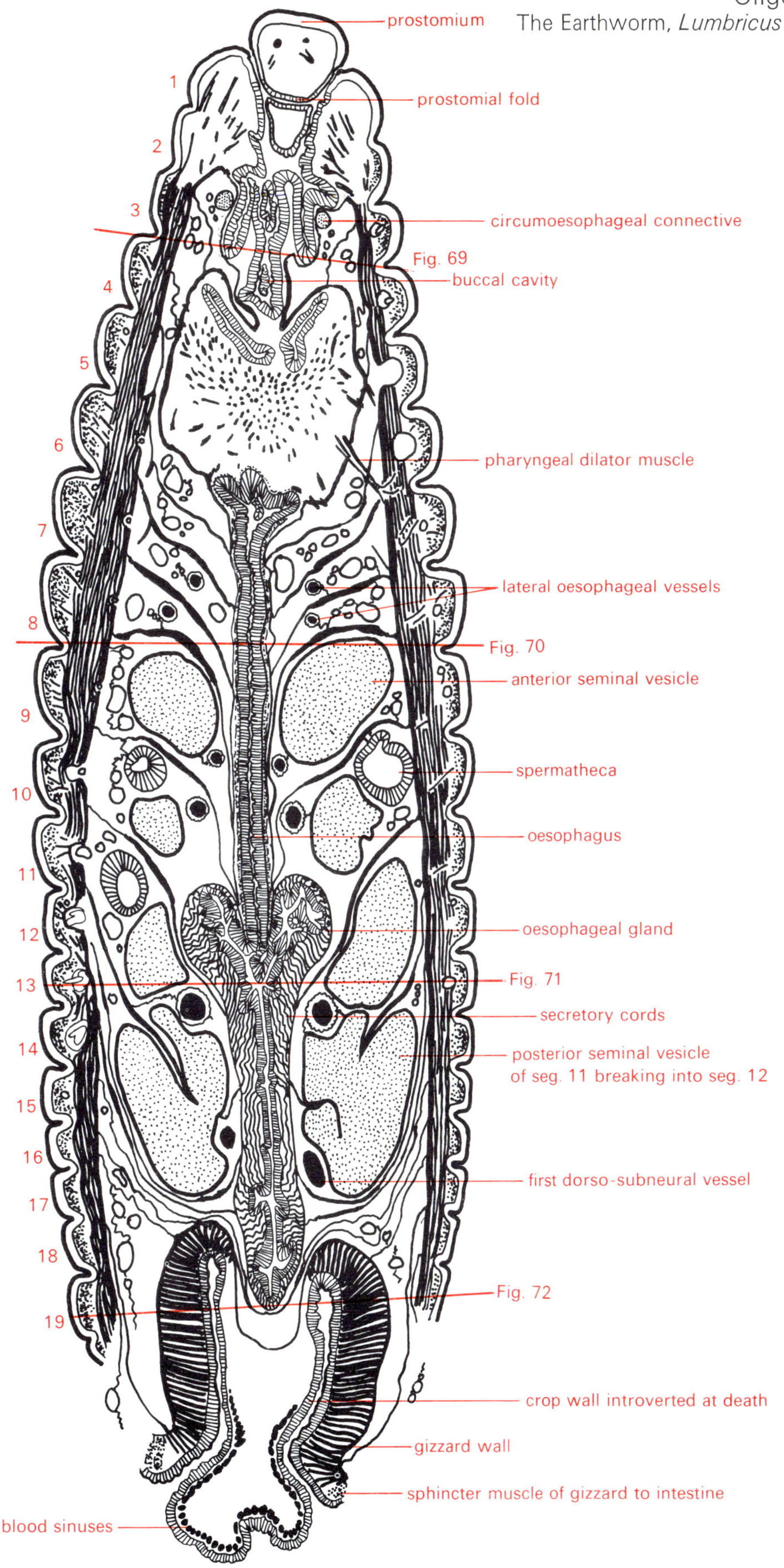

prostomium
prostomial fold
1
2
3
circumoesophageal connective
Fig. 69
buccal cavity
4
5
6
pharyngeal dilator muscle
7
lateral oesophageal vessels
8
Fig. 70
anterior seminal vesicle
9
spermatheca
10
oesophagus
11
oesophageal gland
12
13
Fig. 71
secretory cords
14
posterior seminal vesicle
of seg. 11 breaking into seg. 12
15
16
first dorso-subneural vessel
17
18
Fig. 72
19
crop wall introverted at death
gizzard wall
sphincter muscle of gizzard to intestine
blood sinuses

Fig. 68. H.L.S. of body as far as the gizzard. The positions of the sections in Figs. 69–71 are indicated

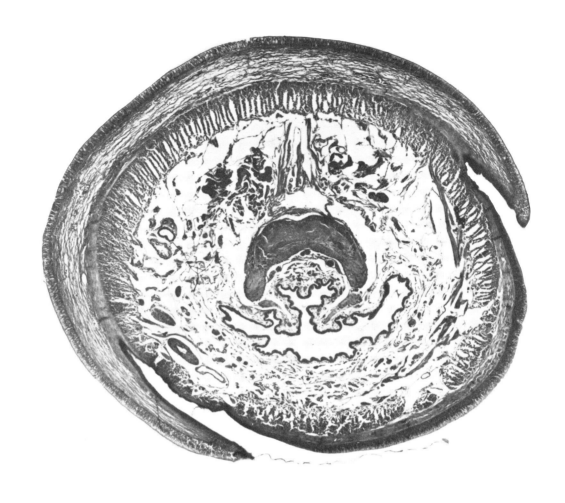

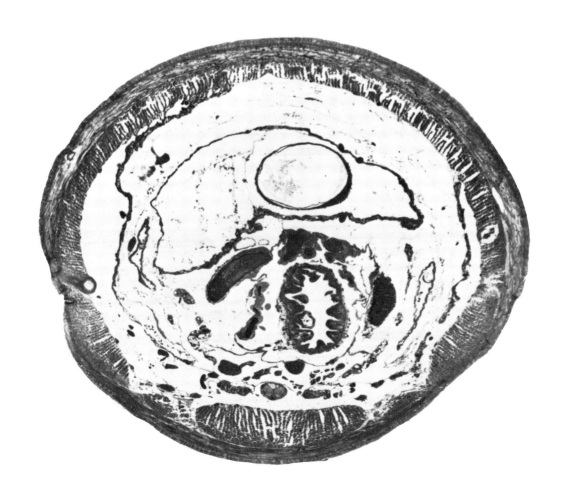

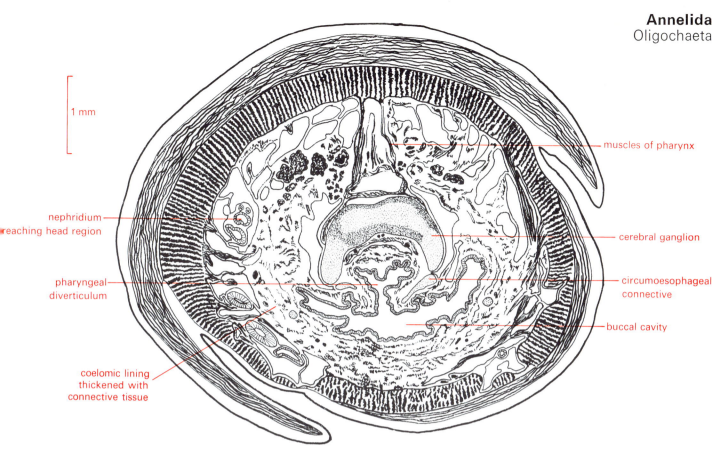

1 mm

muscles of pharynx

nephridium reaching head region

cerebral ganglion

pharyngeal diverticulum

circumoesophageal connective

buccal cavity

coelomic lining thickened with connective tissue

Fig. 69. *Lumbricus* – T.S. region of cerebral ganglion

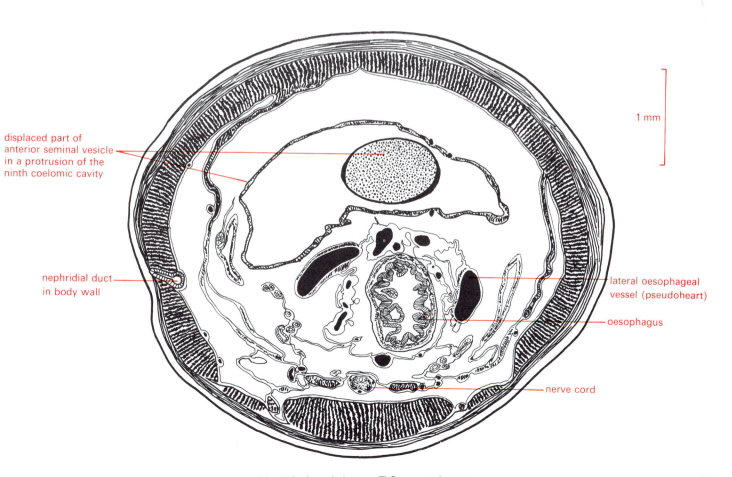

1 mm

displaced part of anterior seminal vesicle in a protrusion of the ninth coelomic cavity

nephridial duct in body wall

lateral oesophageal vessel (pseudoheart)

oesophagus

nerve cord

Fig. 70. *Lumbricus* – T.S. oesophagus

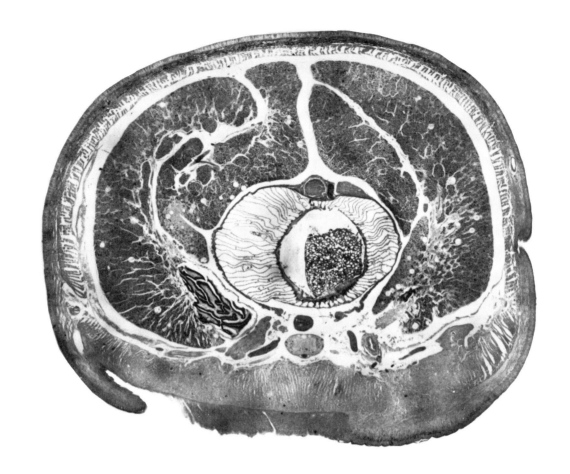

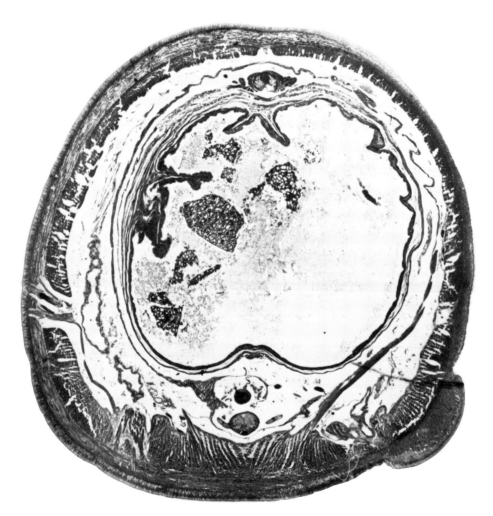

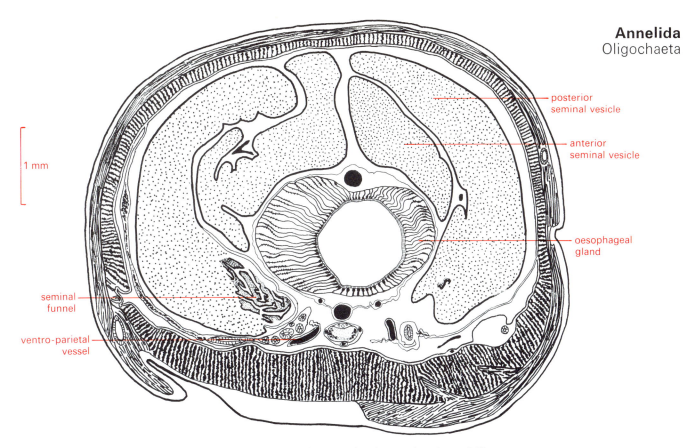

posterior
seminal vesicle

anterior
seminal vesicle

oesophageal
gland

1 mm

seminal
funnel

ventro-parietal
vessel

Fig. 71. *Lumbricus* – T.S. reproductive region (seg. 13)

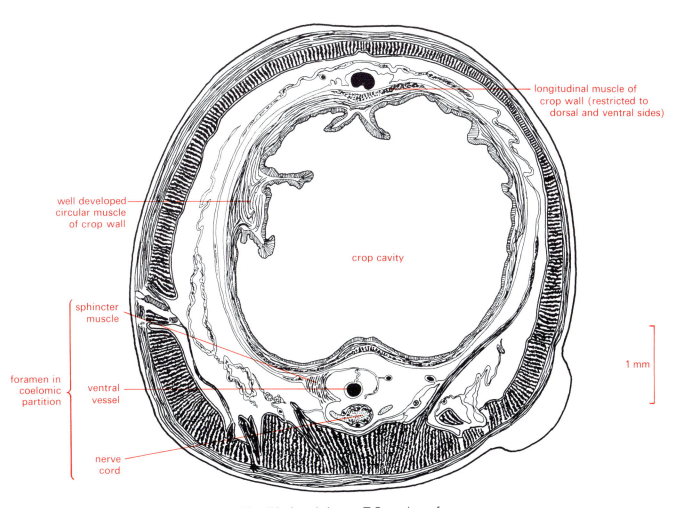

longitudinal muscle of
crop wall (restricted to
dorsal and ventral sides)

well developed
circular muscle
of crop wall

crop cavity

sphincter
muscle

1 mm

foramen in
coelomic
partition

ventral
vessel

nerve
cord

Fig. 72. *Lumbricus* – T.S. region of crop

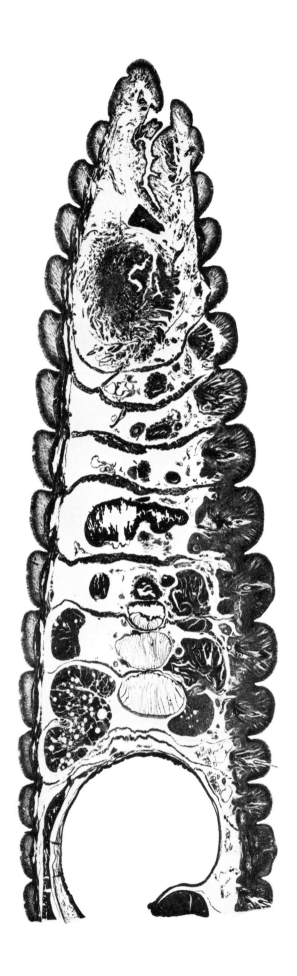

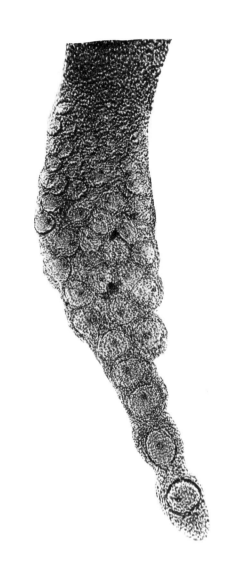

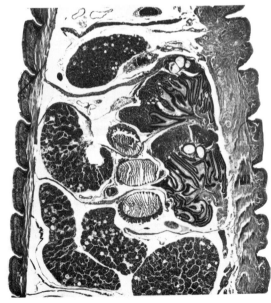

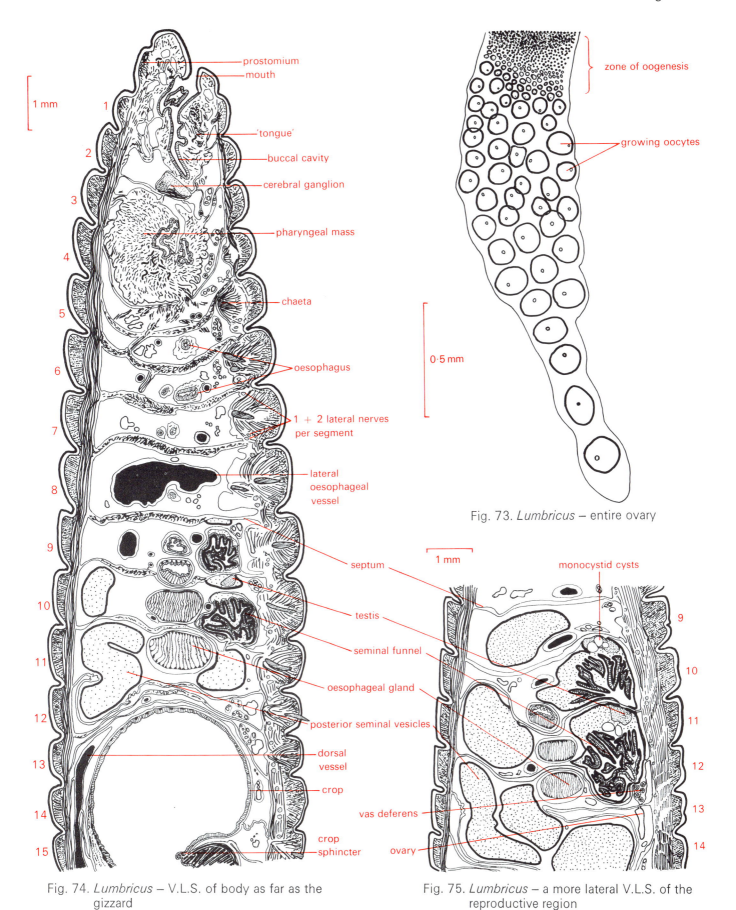

prostomium
mouth
'tongue'
buccal cavity
cerebral ganglion
pharyngeal mass
chaeta
oesophagus
1 + 2 lateral nerves per segment
lateral oesophageal vessel
septum
testis
seminal funnel
oesophageal gland
posterior seminal vesicles
dorsal vessel
crop
crop sphincter

1 mm

Fig. 74. *Lumbricus* – V.L.S. of body as far as the gizzard

zone of oogenesis
growing oocytes

0·5 mm

Fig. 73. *Lumbricus* – entire ovary

1 mm

monocystid cysts
vas deferens
ovary

Fig. 75. *Lumbricus* – a more lateral V.L.S. of the reproductive region

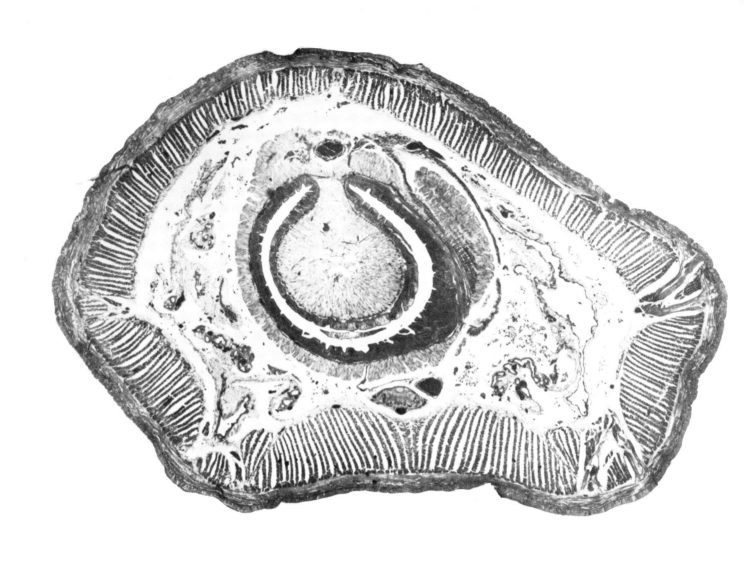

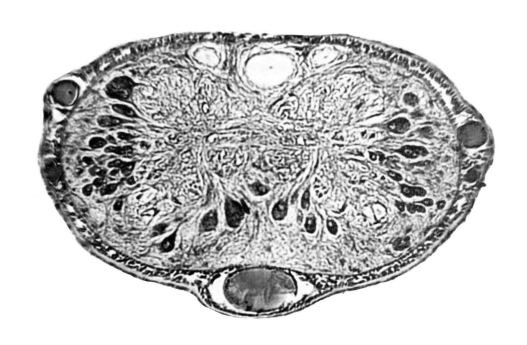

54

Annelida
Oligochaeta

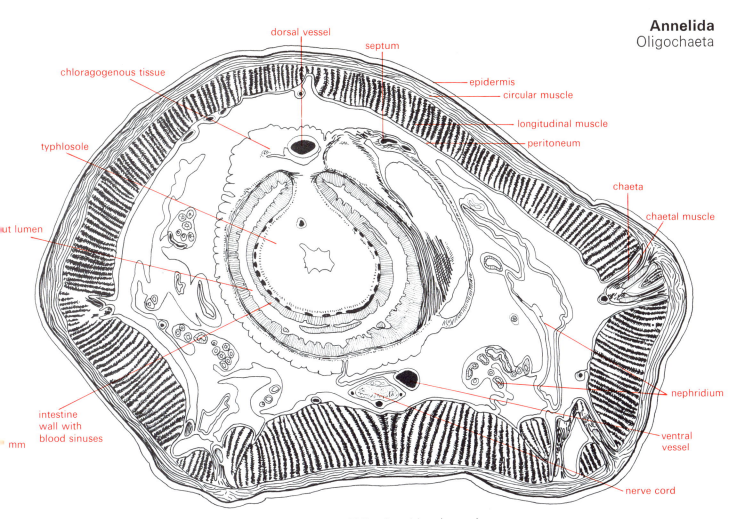

Fig. 76. *Lumbricus* – T.S. of typhlosolar region

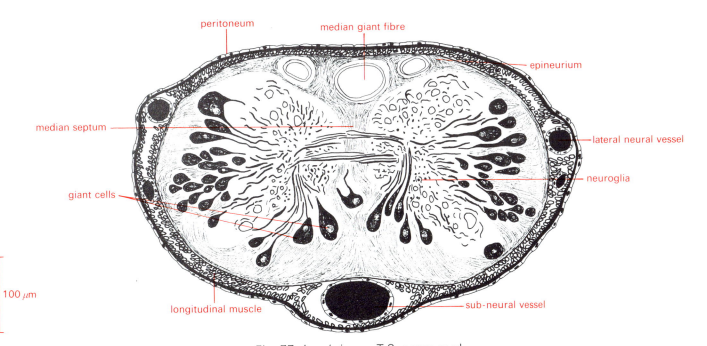

Fig. 77. *Lumbricus* – T.S. nerve cord

55

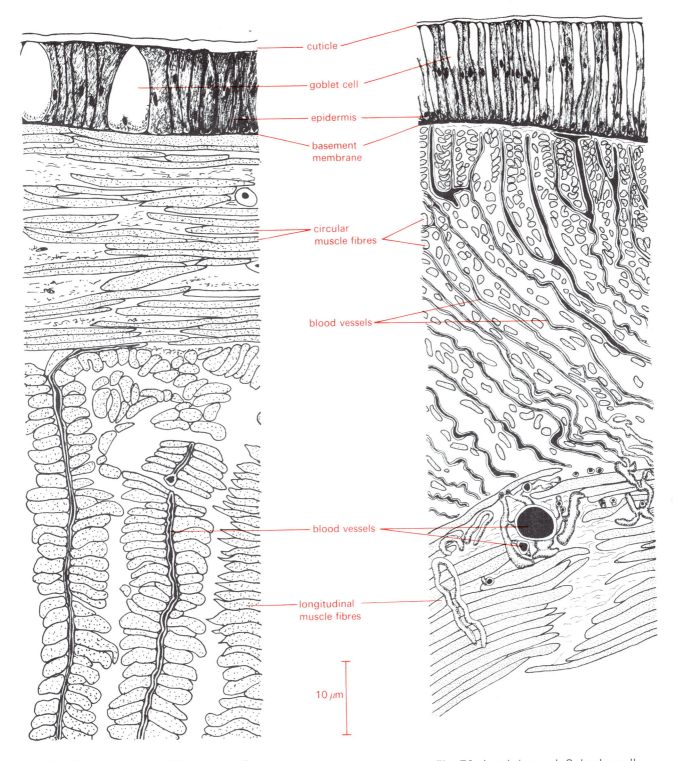

cuticle

goblet cell

epidermis

basement
membrane

circular
muscle fibres

blood vessels

blood vessels

longitudinal
muscle fibres

10 μm

Fig. 78. *Lumbricus* – T.S. body wall

Fig. 79. *Lumbricus* – L.S. body wall

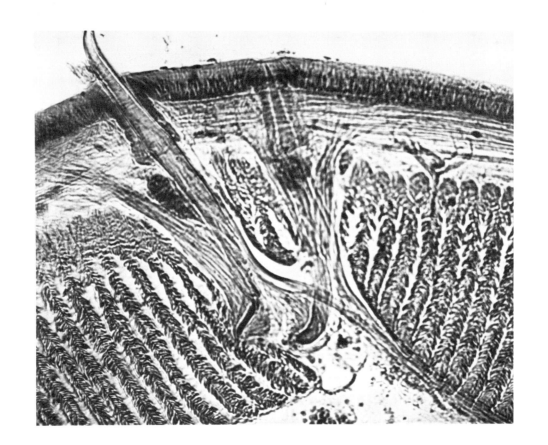

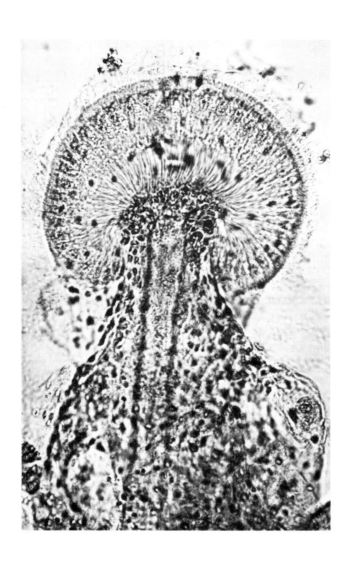

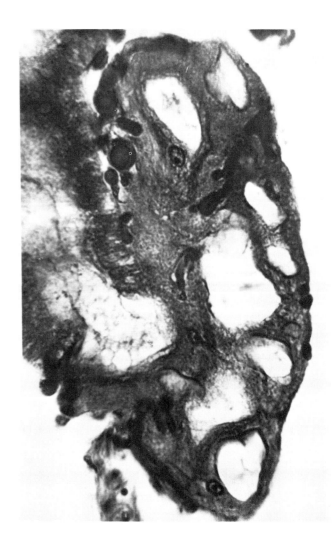

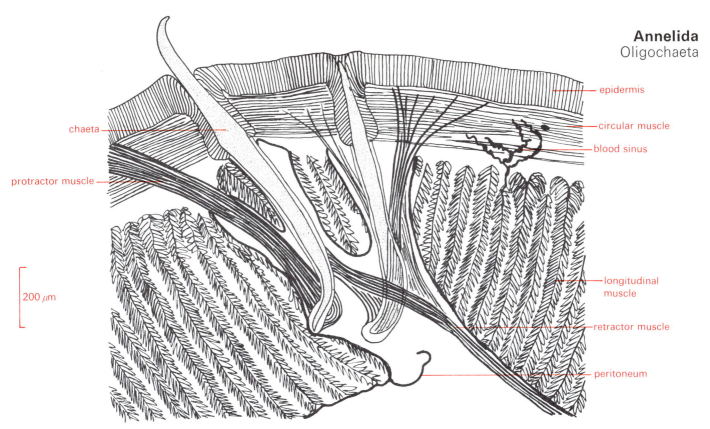

epidermis

circular muscle

blood sinus

chaeta

protractor muscle

longitudinal muscle

retractor muscle

peritoneum

200 μm

Fig. 80. *Lumbricus* – thick T.S. body wall to show insertion of chaetae

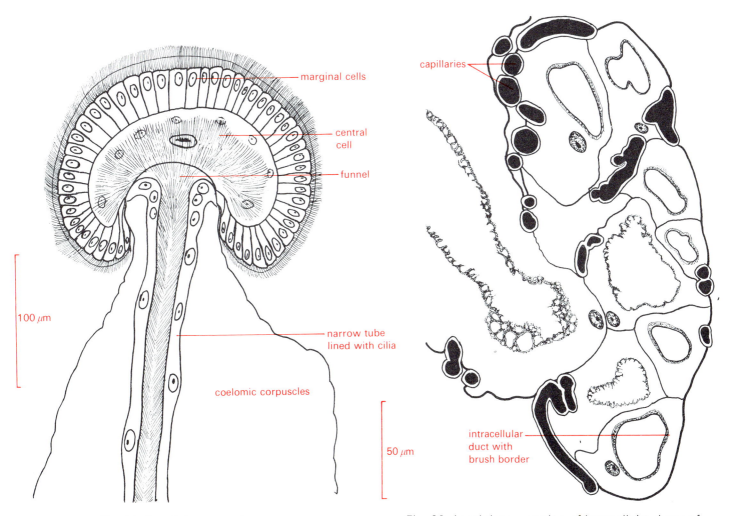

marginal cells

central cell

funnel

narrow tube lined with cilia

coelomic corpuscles

100 μm

capillaries

intracellular duct with brush border

50 μm

Fig. 81. *Lumbricus* – nephrostome

Fig. 82. *Lumbricus* – section of intracellular ducts of nephridium

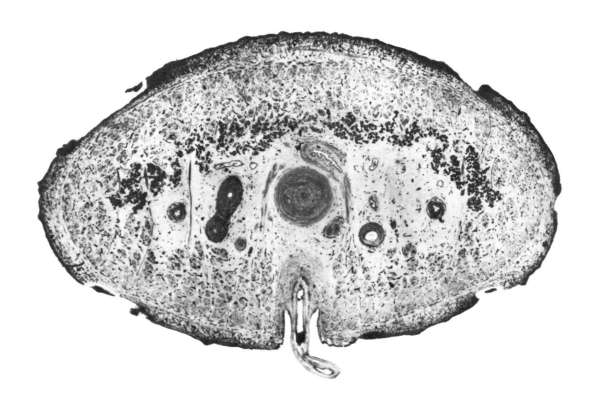

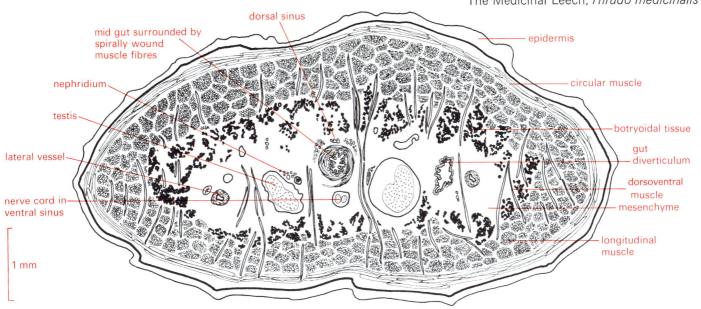

dorsal sinus

mid gut surrounded by
spirally wound
muscle fibres

nephridium

testis

lateral vessel

nerve cord in
ventral sinus

epidermis

circular muscle

botryoidal tissue

gut
diverticulum

dorsoventral
muscle

mesenchyme

longitudinal
muscle

1 mm

Fig. 83. T.S. body I

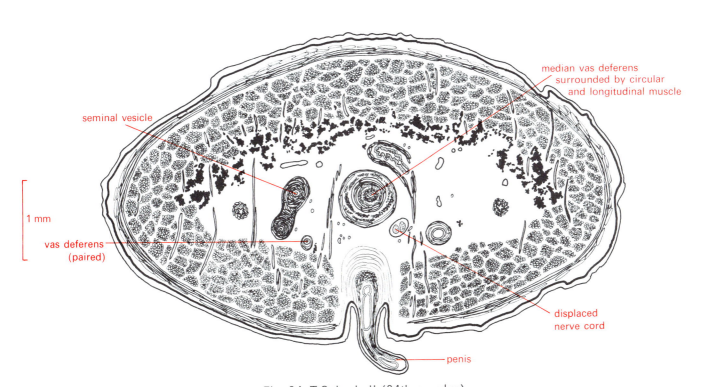

median vas deferens
surrounded by circular
and longitudinal muscle

seminal vesicle

1 mm

vas deferens
(paired)

displaced
nerve cord

penis

Fig. 84. T.S. body II (24th annulus)

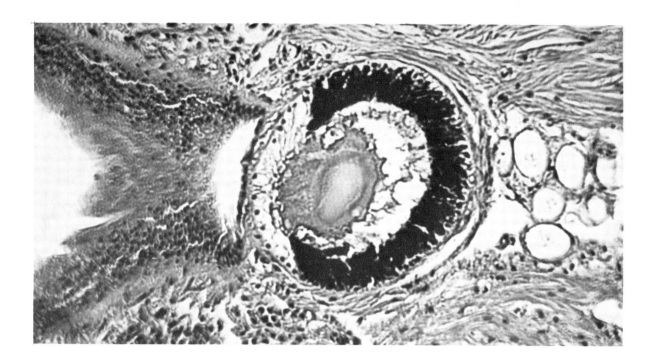

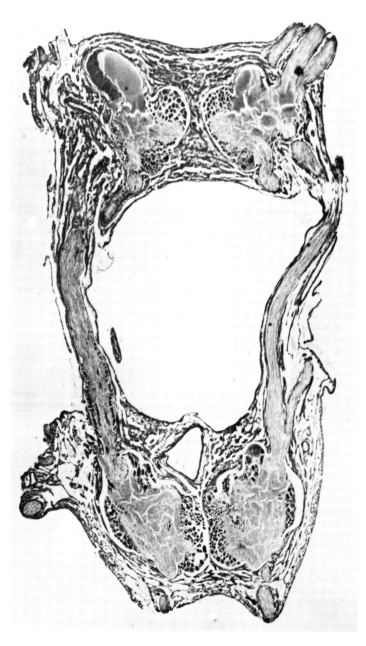

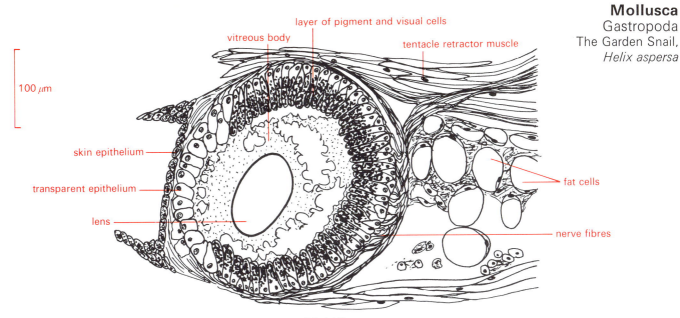

100 μm

layer of pigment and visual cells

vitreous body

tentacle retractor muscle

skin epithelium

transparent epithelium

lens

fat cells

nerve fibres

Fig. 85. V.S. eye

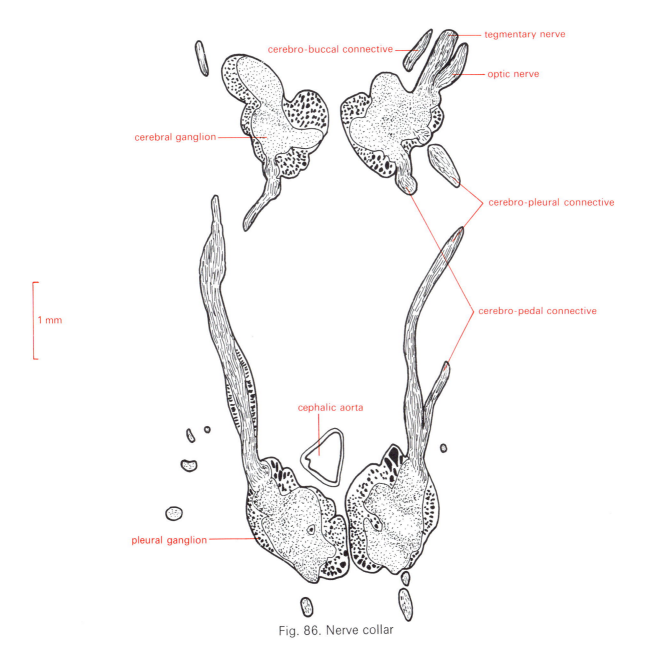

tegmentary nerve

cerebro-buccal connective

optic nerve

cerebral ganglion

cerebro-pleural connective

cerebro-pedal connective

1 mm

cephalic aorta

pleural ganglion

Fig. 86. Nerve collar

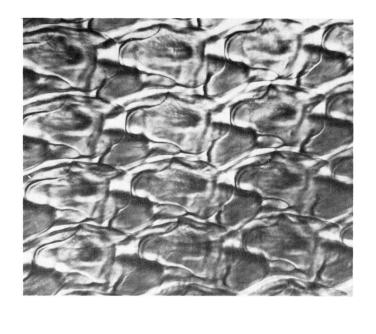

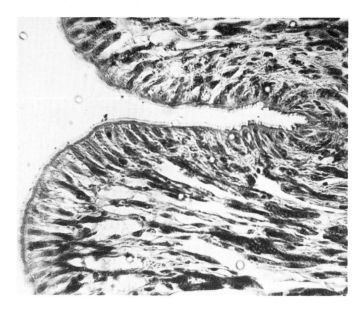

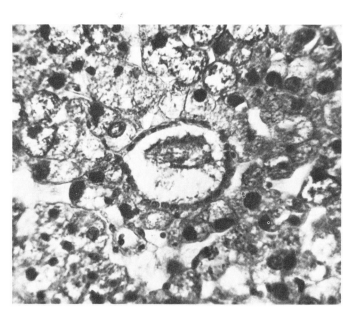

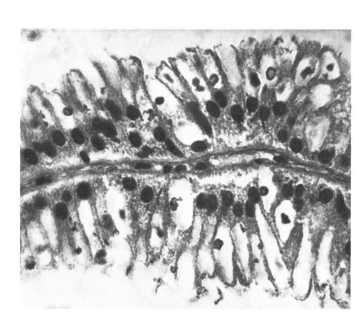

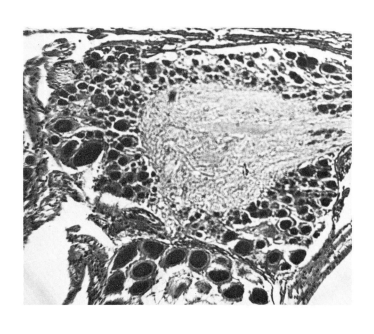

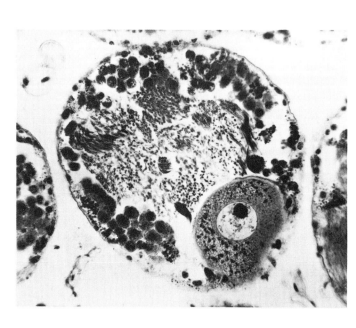

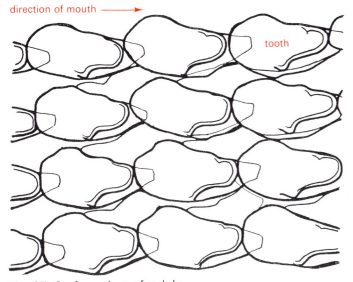

direction of mouth

tooth

20 μm

Fig. 87. Surface view of radula

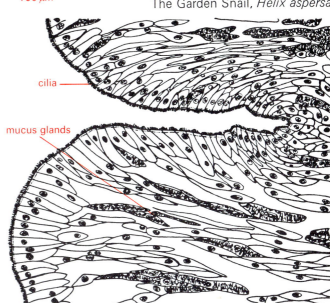

100 μm

cilia

mucus glands

Fig. 88. V.S. foot

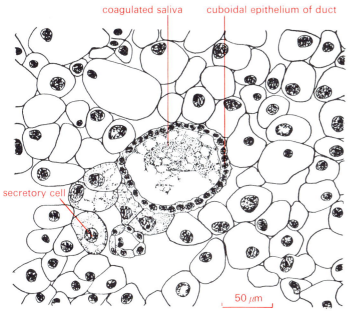

coagulated saliva   cuboidal epithelium of duct

secretory cell

50 μm

Fig. 89. T.S. salivary gland

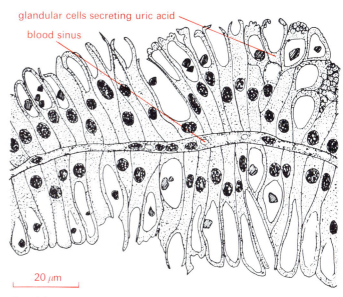

glandular cells secreting uric acid

blood sinus

20 μm

Fig. 90. V.S. kidney

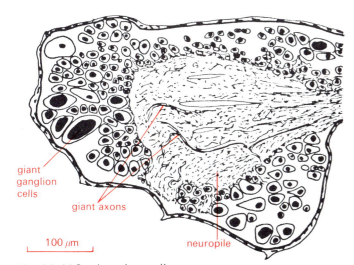

giant ganglion cells

giant axons

neuropile

100 μm

Fig. 91. V.S. pleural ganglion

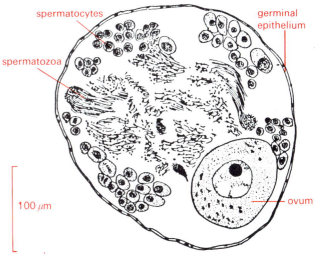

spermatocytes

germinal epithelium

spermatozoa

ovum

100 μm

Fig. 92. T.S. follicle of hermaphrodite gland

65

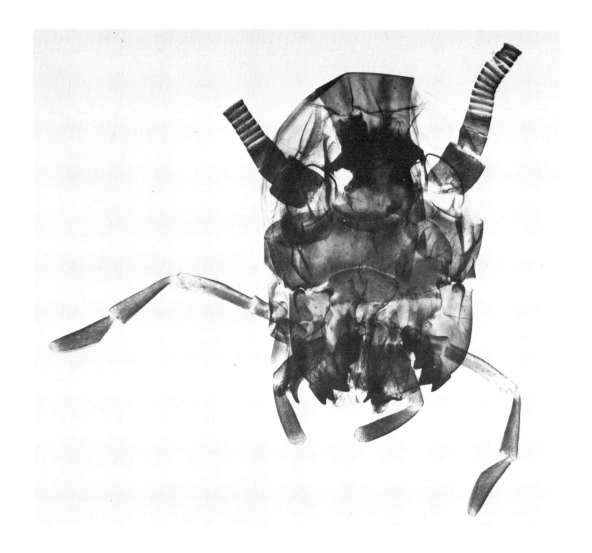

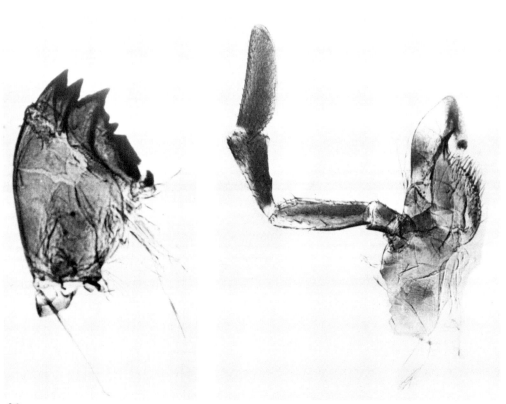

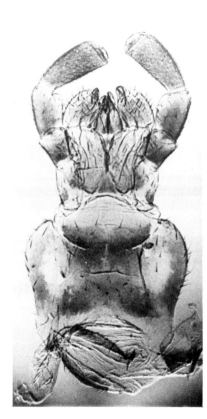

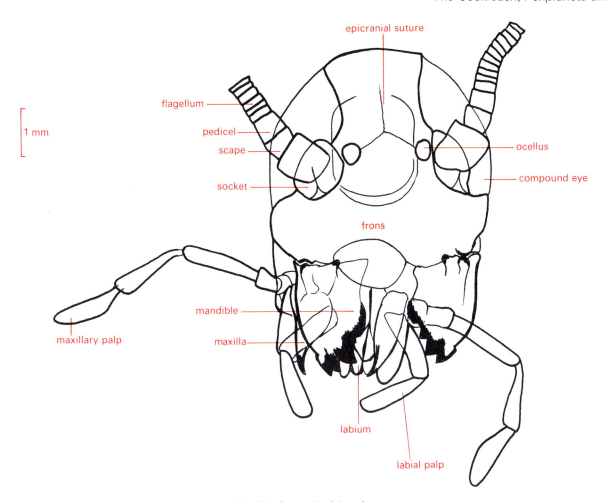

Fig. 93. Squashed head

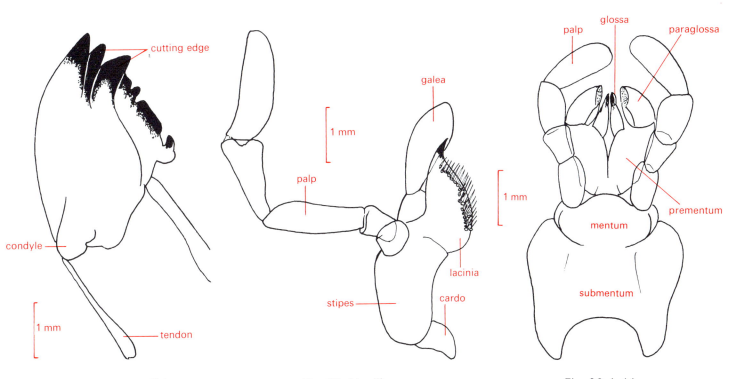

Fig. 94. Mandible   Fig. 95. Maxilla   Fig. 96. Labium

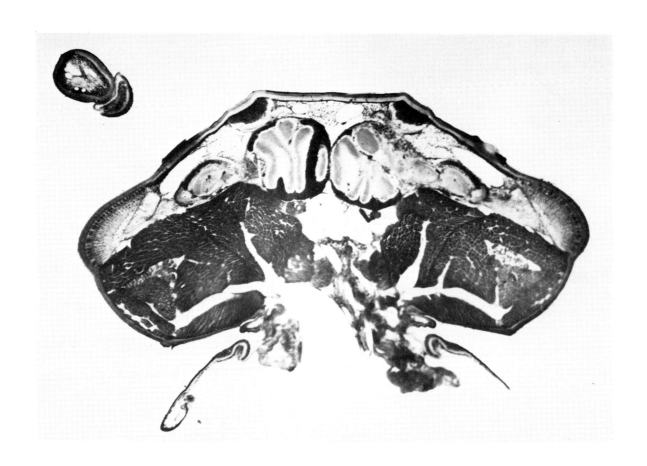

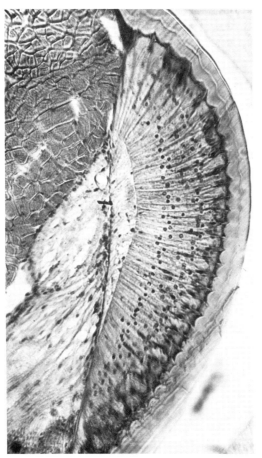

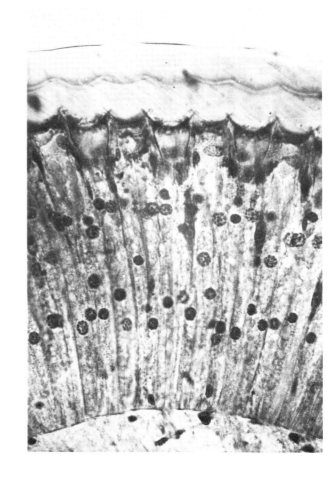

68

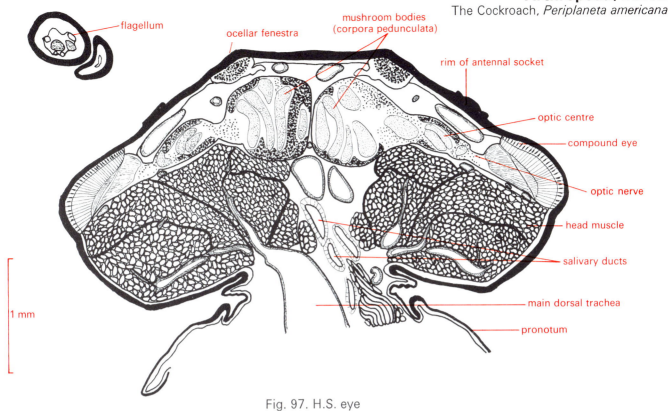

flagellum

ocellar fenestra

mushroom bodies
(corpora pedunculata)

rim of antennal socket

optic centre

compound eye

optic nerve

head muscle

salivary ducts

main dorsal trachea

pronotum

1 mm

Fig. 97. H.S. eye

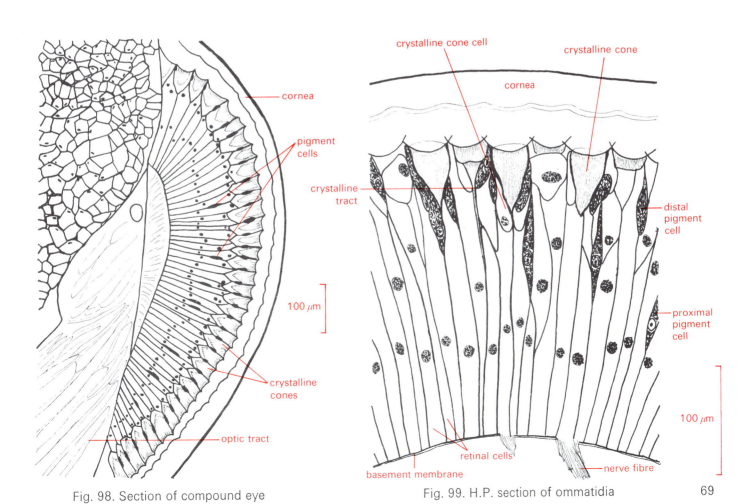

cornea

pigment
cells

crystalline
cones

optic tract

100 μm

crystalline cone cell

crystalline cone

cornea

crystalline
tract

distal
pigment
cell

proximal
pigment
cell

retinal cells

basement membrane

nerve fibre

100 μm

Fig. 98. Section of compound eye

Fig. 99. H.P. section of ommatidia

69

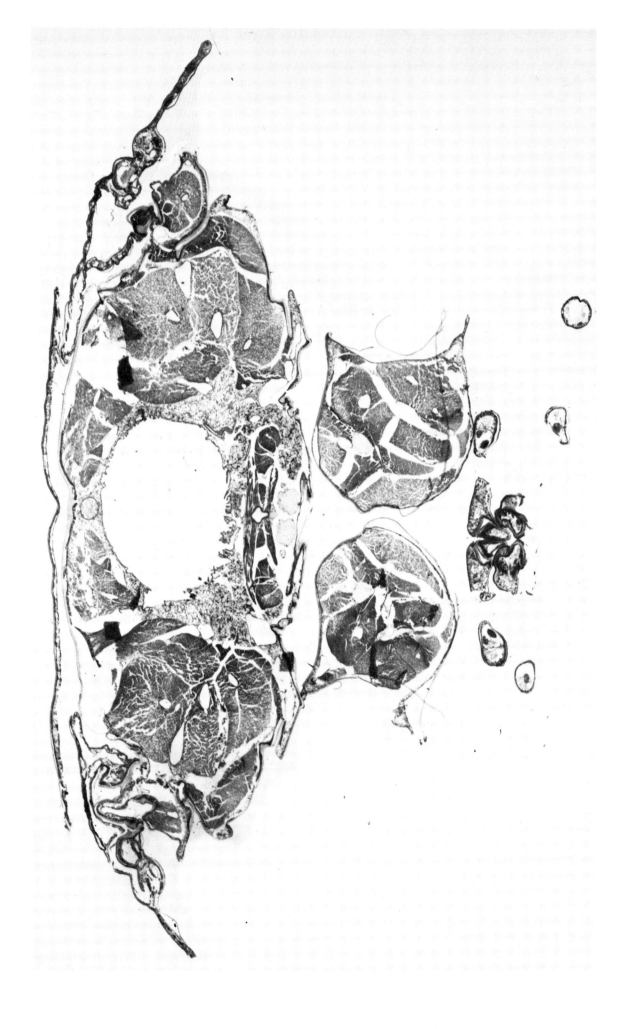

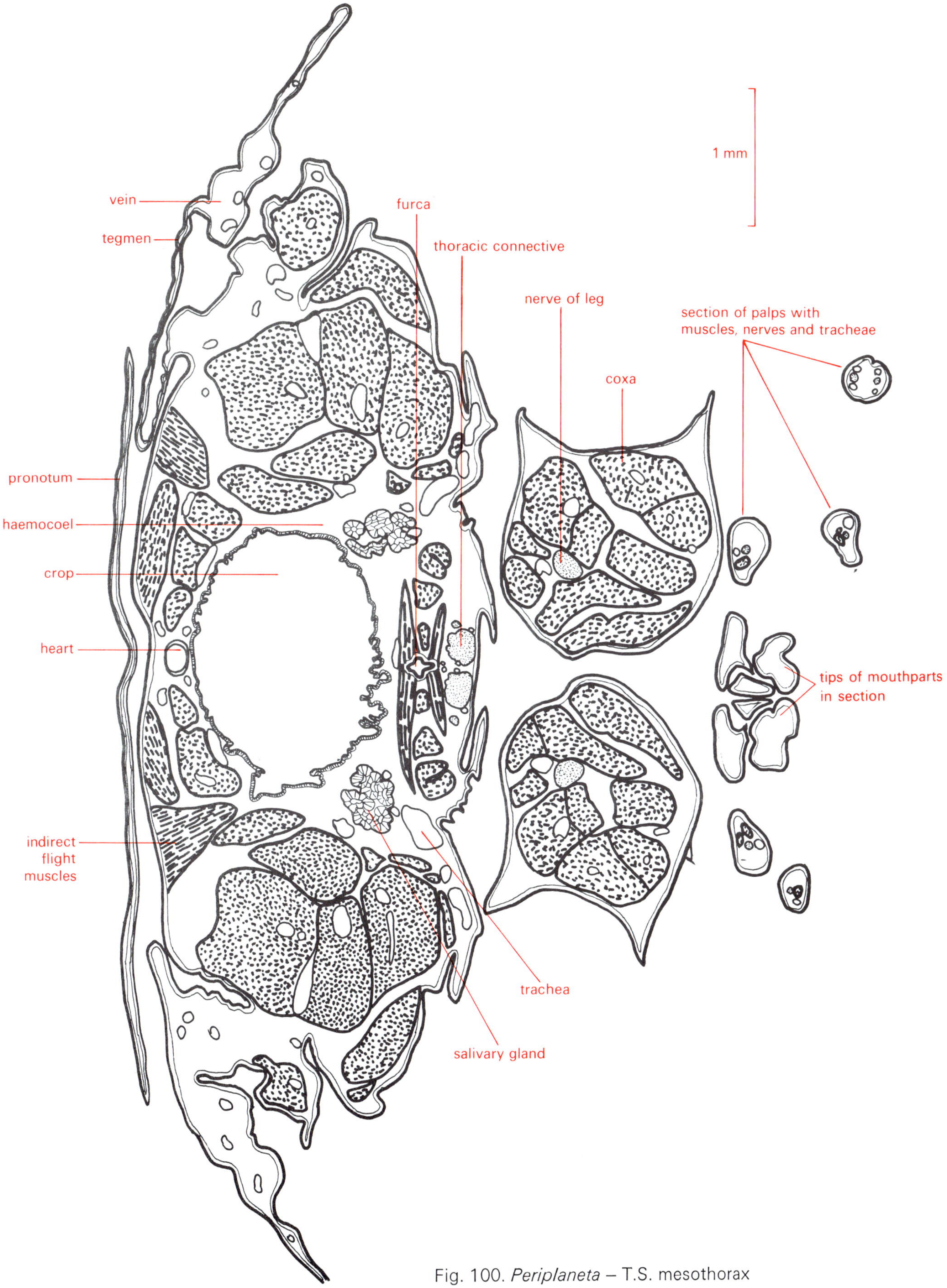

1 mm

vein

tegmen

furca

thoracic connective

nerve of leg

section of palps with
muscles, nerves and tracheae

coxa

pronotum

haemocoel

crop

heart

tips of mouthparts
in section

indirect
flight
muscles

trachea

salivary gland

Fig. 100. *Periplaneta* – T.S. mesothorax

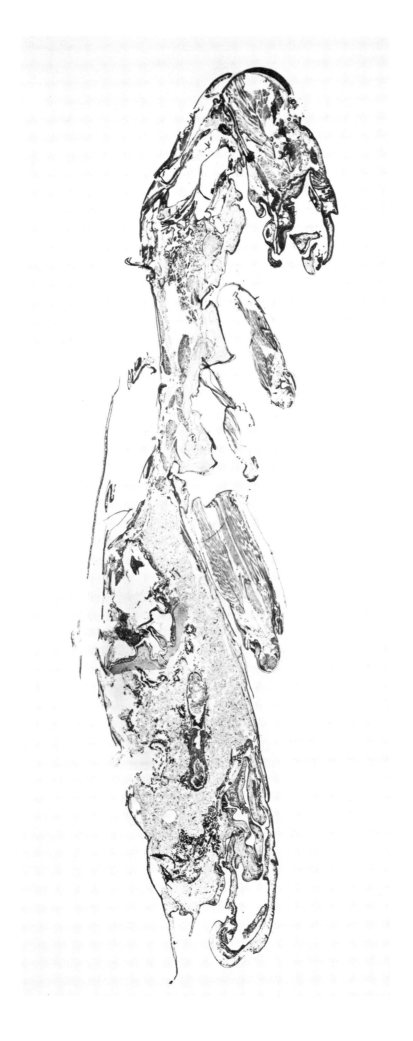

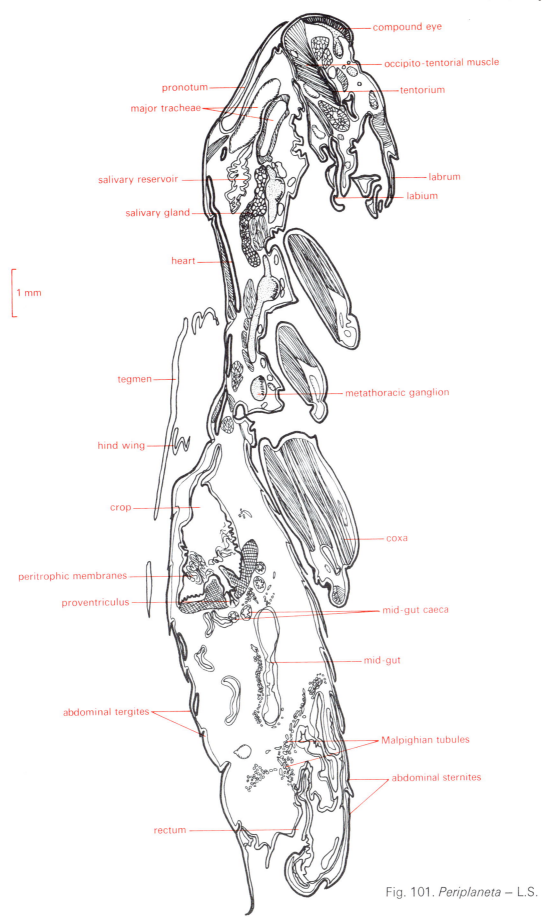

compound eye

occipito-tentorial muscle

pronotum

tentorium

major tracheae

labrum

labium

salivary reservoir

salivary gland

heart

1 mm

tegmen

metathoracic ganglion

hind wing

crop

coxa

peritrophic membranes

proventriculus

mid-gut caeca

mid-gut

abdominal tergites

Malpighian tubules

abdominal sternites

rectum

Fig. 101. *Periplaneta* – L.S.

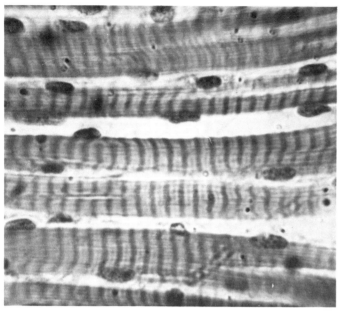
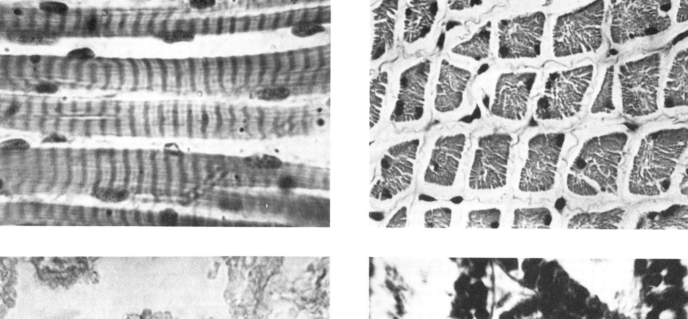
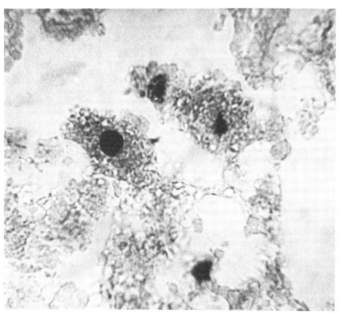
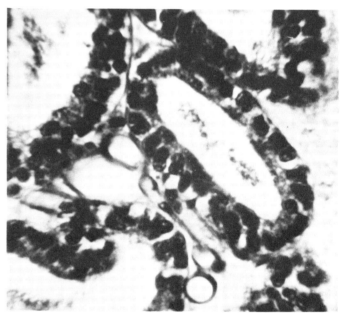
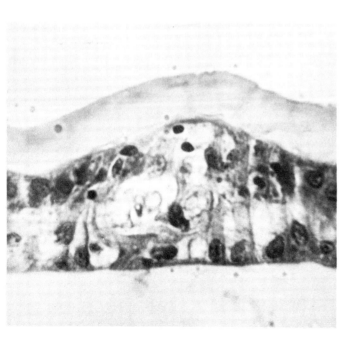
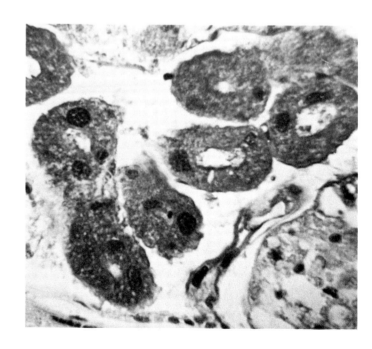

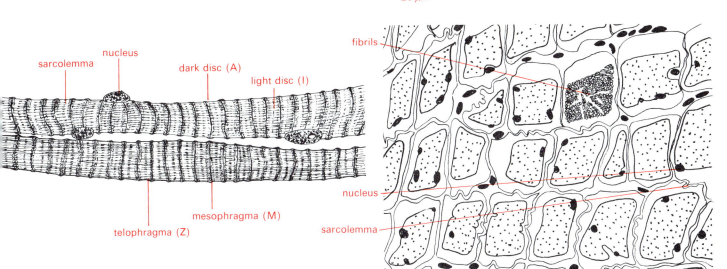

Fig. 102. Muscle fibres – surface view

Fig. 103. Muscle fibres – T.S. (shrunk in fixation)

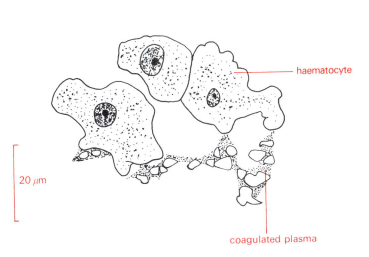

Fig. 104. Blood

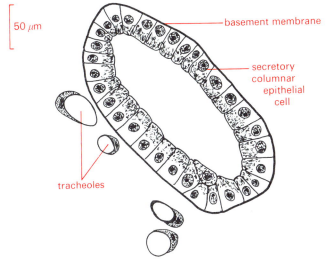

Fig. 105. Mid-gut caeca

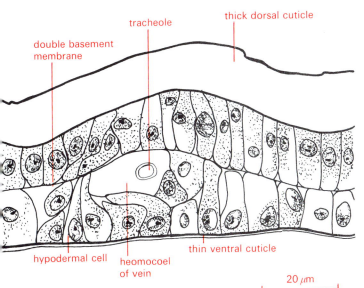

Fig. 106. T.S. tegmen

Fig. 107. T.S. Malpighian tubules

75

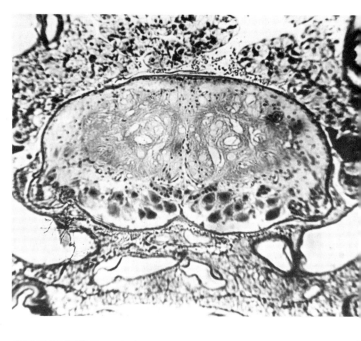
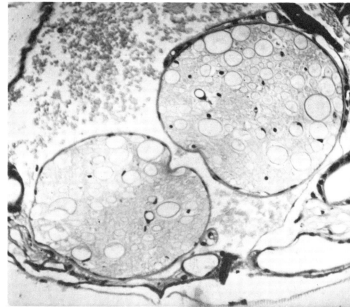
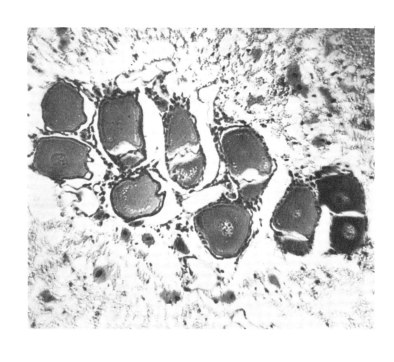
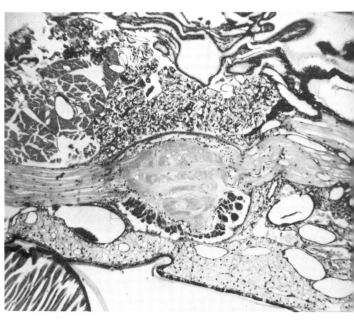

Fig. 108. Salivary apparatus

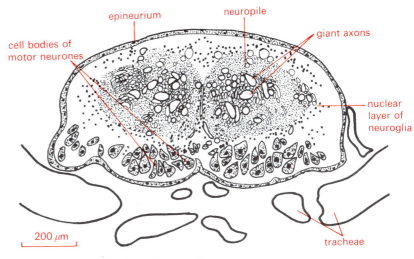

Fig. 110. T.S. thoracic ganglion

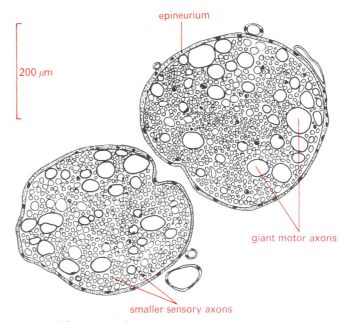

Fig. 111. T.S. connectives

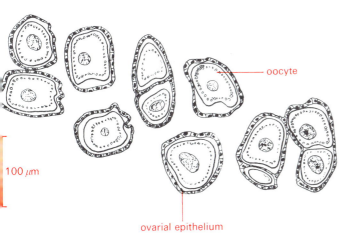

Fig. 109. Section of ovarioles (shrunk in fixation)

Fig. 112. L.S. thoracic ganglion

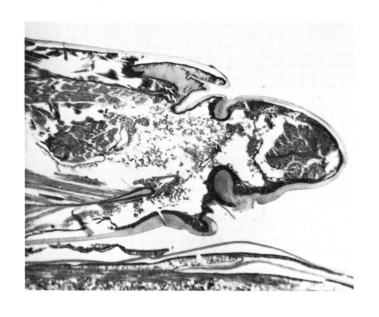
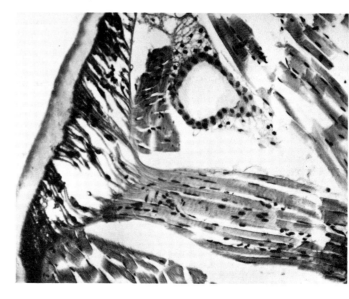
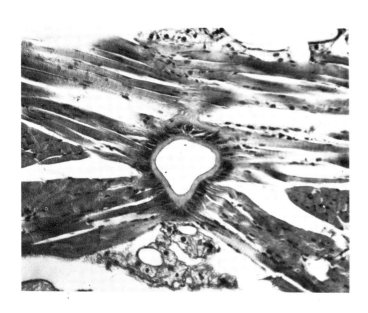
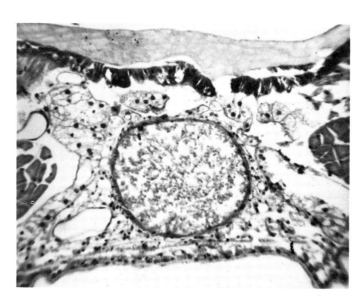
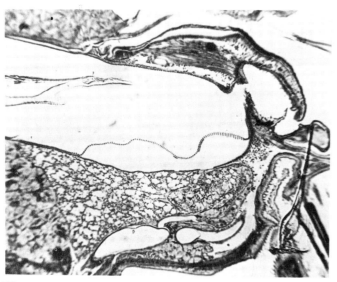
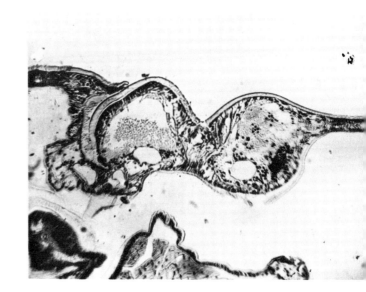

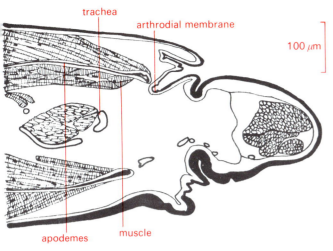

Fig. 113. L.S. coxo-femoral joint

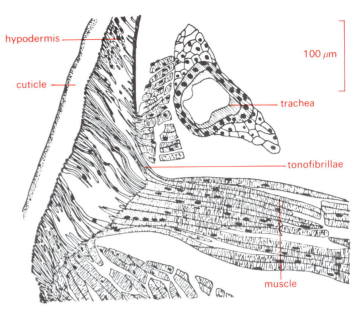

Fig. 114. T.S. body showing muscle attachment

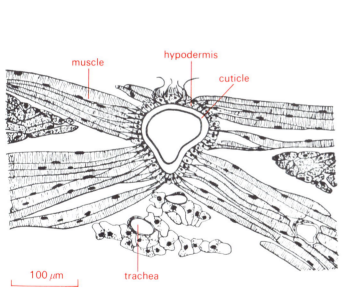

Fig. 115. T.S. thoracic furca

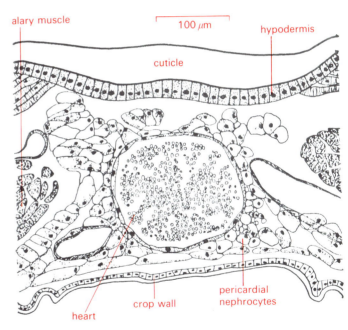

Fig. 116. T.S. heart

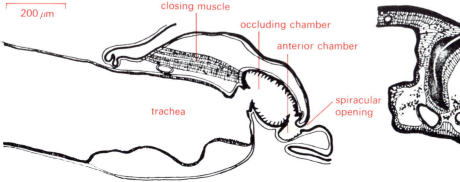

Fig. 117. L.S. thoracic spiracle

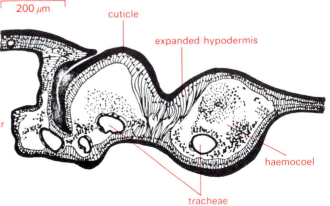

Fig. 118. T.S. leading edge of tegmen

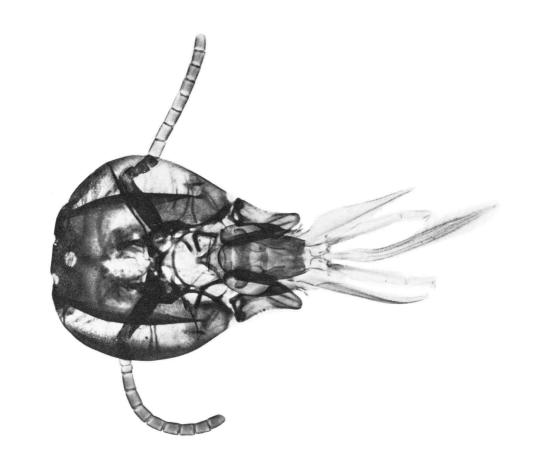

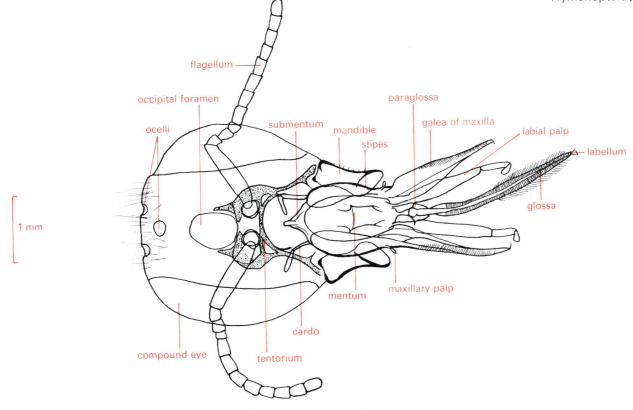

Fig. 119. *Apis* – squashed head of adult worker

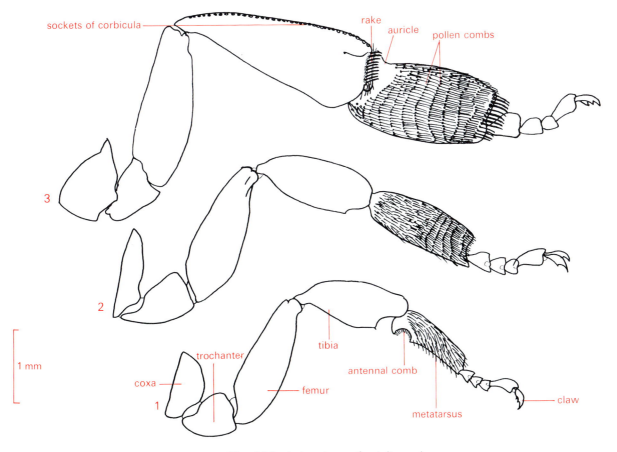

Fig. 120. *Apis* – legs of adult worker

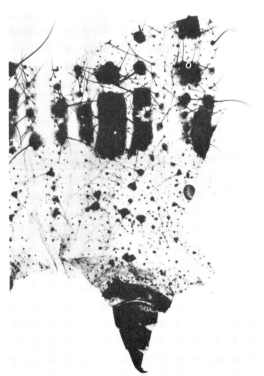 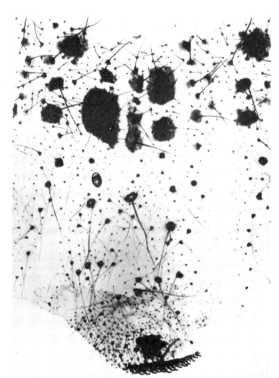

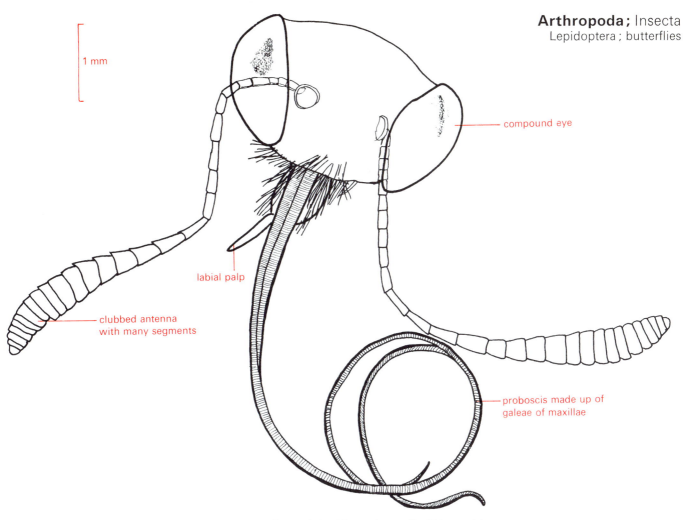

Fig. 121. Head of an adult skipper butterfly (*Hesperis*)

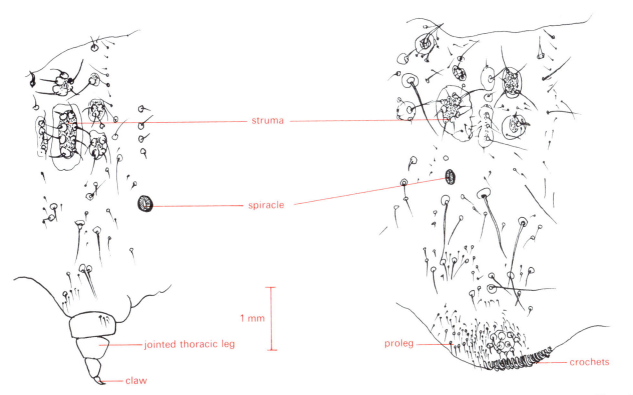

Fig. 122. Thoracic segment of the caterpillar of the cabbage white butterfly (*Pieris brassicae*)

Fig. 123. Abdominal segment of the caterpillar of the cabbage white butterfly

83

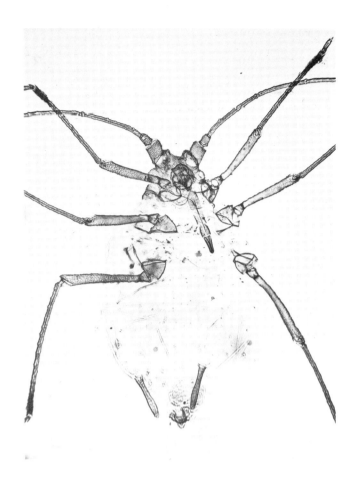

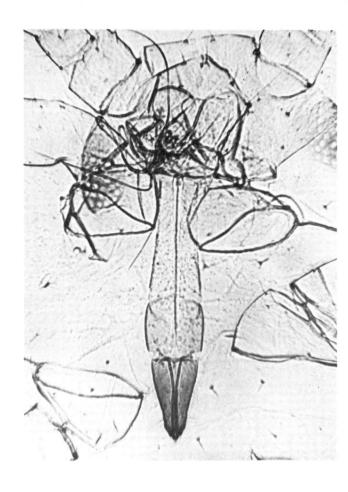

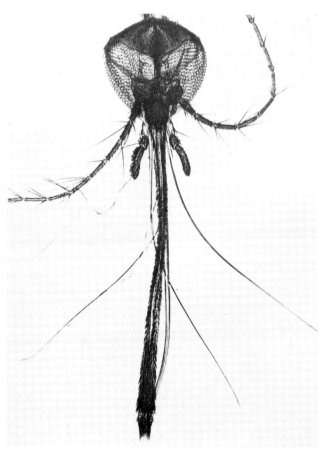

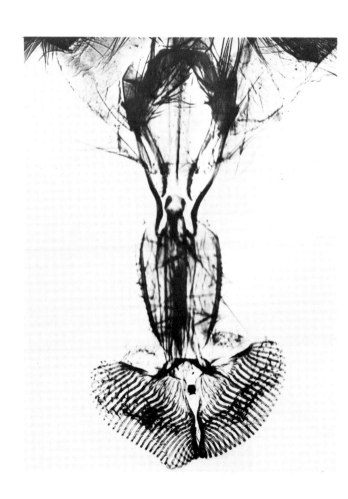

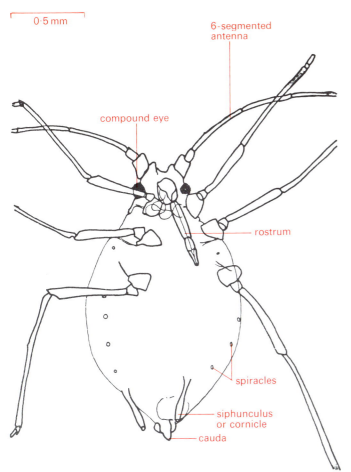

Fig. 124. *Myzus* – adult apterous aphid

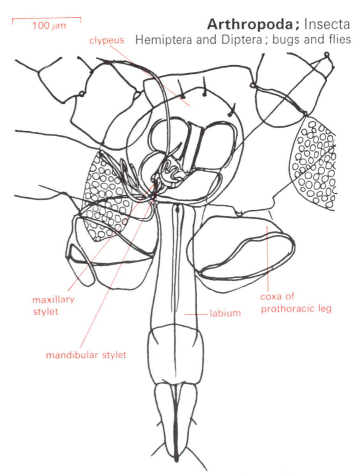

Fig. 125. *Macrosiphum* – mouthparts of an aphid

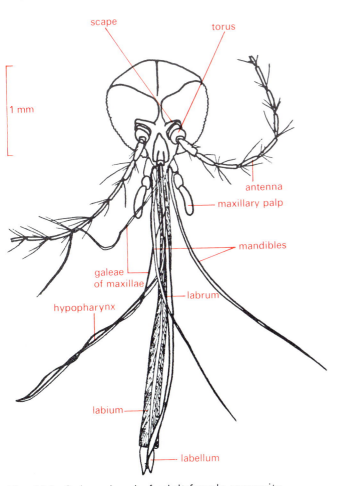

Fig. 126. *Culex* – head of adult female mosquito

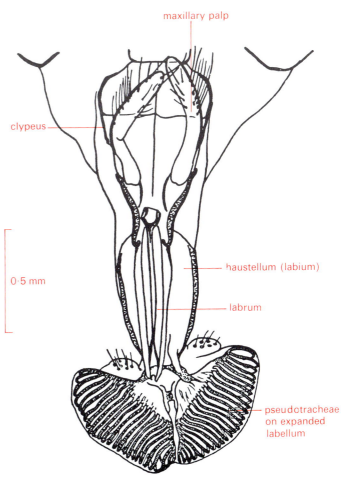

Fig. 127. *Musca* – mouthparts of adult housefly

85

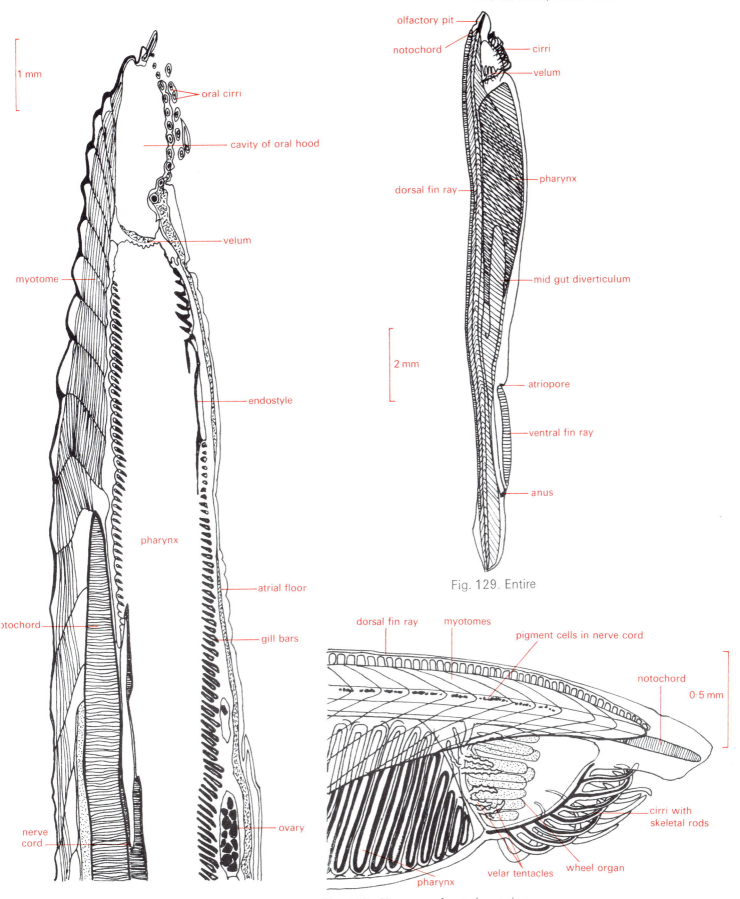

1 mm

oral cirri

cavity of oral hood

velum

myotome

endostyle

pharynx

notochord

atrial floor

gill bars

ovary

nerve cord

Fig. 128. V.L.S. anterior region

olfactory pit

notochord

cirri

velum

pharynx

dorsal fin ray

mid gut diverticulum

2 mm

atriopore

ventral fin ray

anus

Fig. 129. Entire

dorsal fin ray

myotomes

pigment cells in nerve cord

notochord

0·5 mm

cirri with skeletal rods

pharynx

velar tentacles

wheel organ

Fig. 130. Close-up of anterior region

87

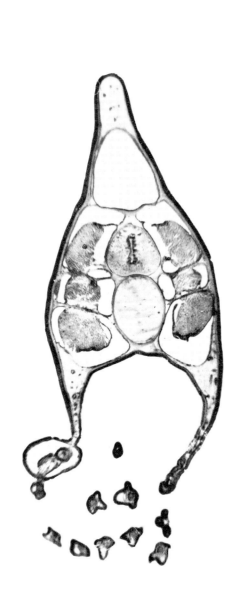

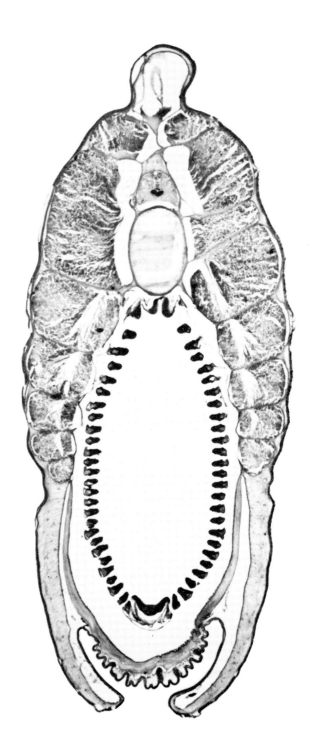

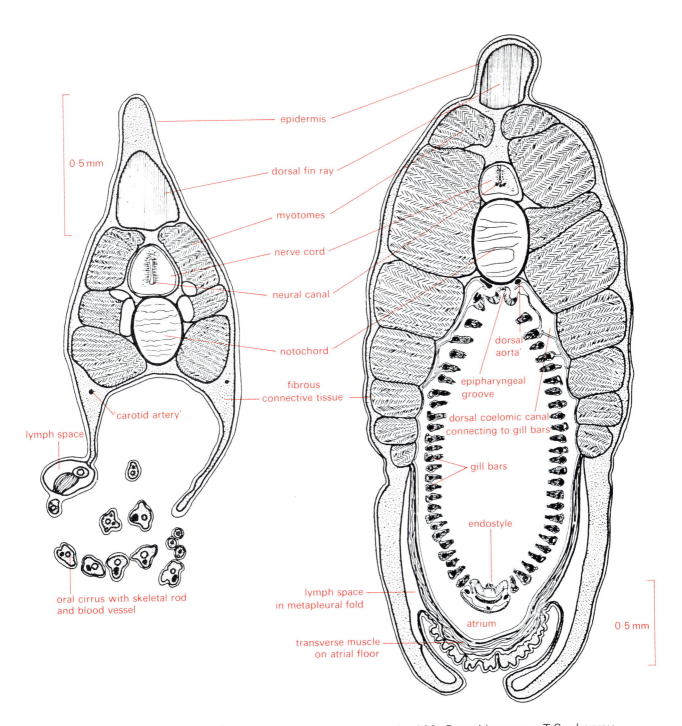

0.5 mm

epidermis

dorsal fin ray

myotomes

nerve cord

neural canal

notochord

fibrous
connective tissue

'carotid artery'

lymph space

oral cirrus with skeletal rod
and blood vessel

dorsal
aorta'

epipharyngeal
groove

dorsal coelomic canal
connecting to gill bars

gill bars

endostyle

lymph space
in metapleural fold

atrium

transverse muscle
on atrial floor

0.5 mm

Fig. 131. *Branchiostoma* – T.S. region of oral hood

Fig. 132. *Branchiostoma* – T.S. pharynx

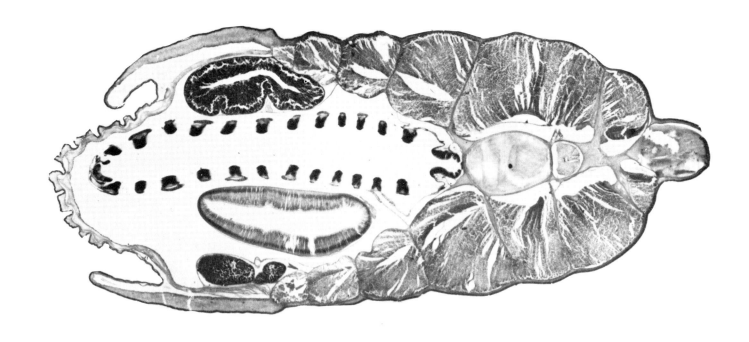

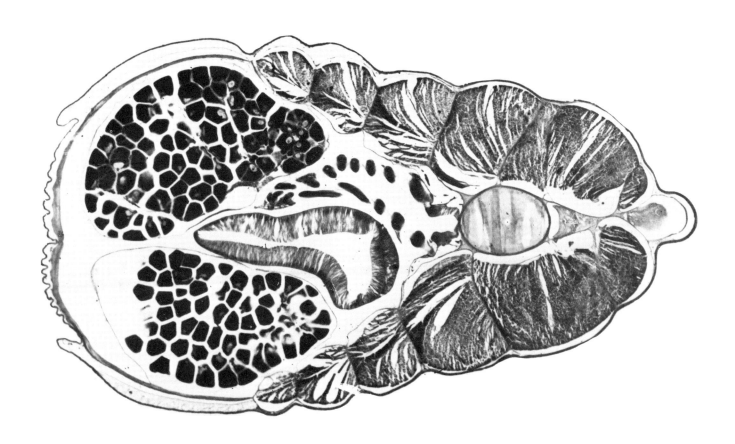

Fig. 133. *Branchiostoma* – T.S. pharynx of male

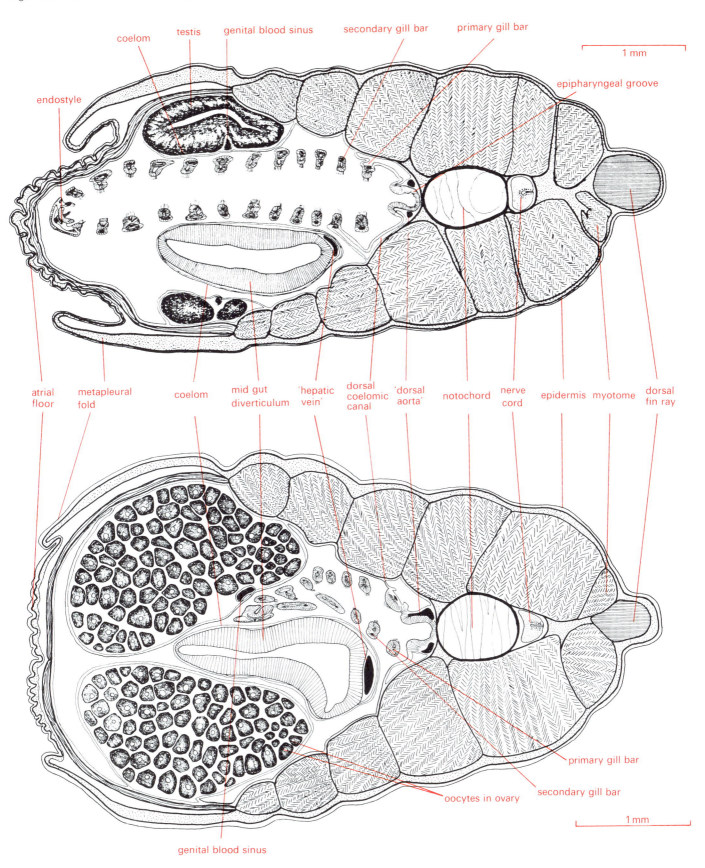

Fig. 134. *Branchiostoma* – T.S. pharynx of female

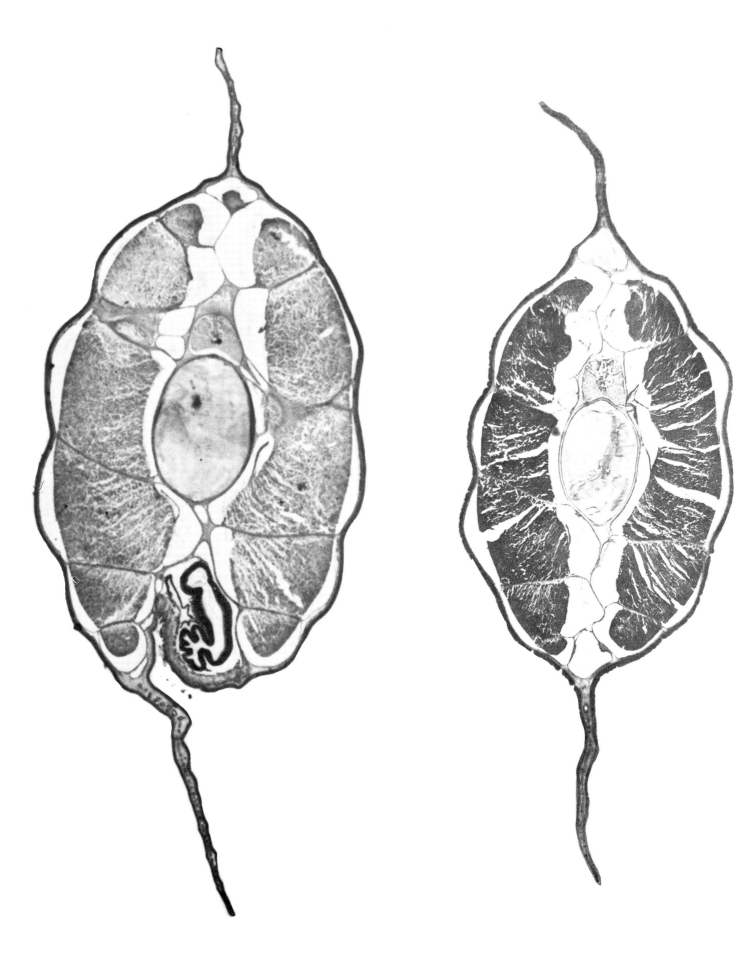

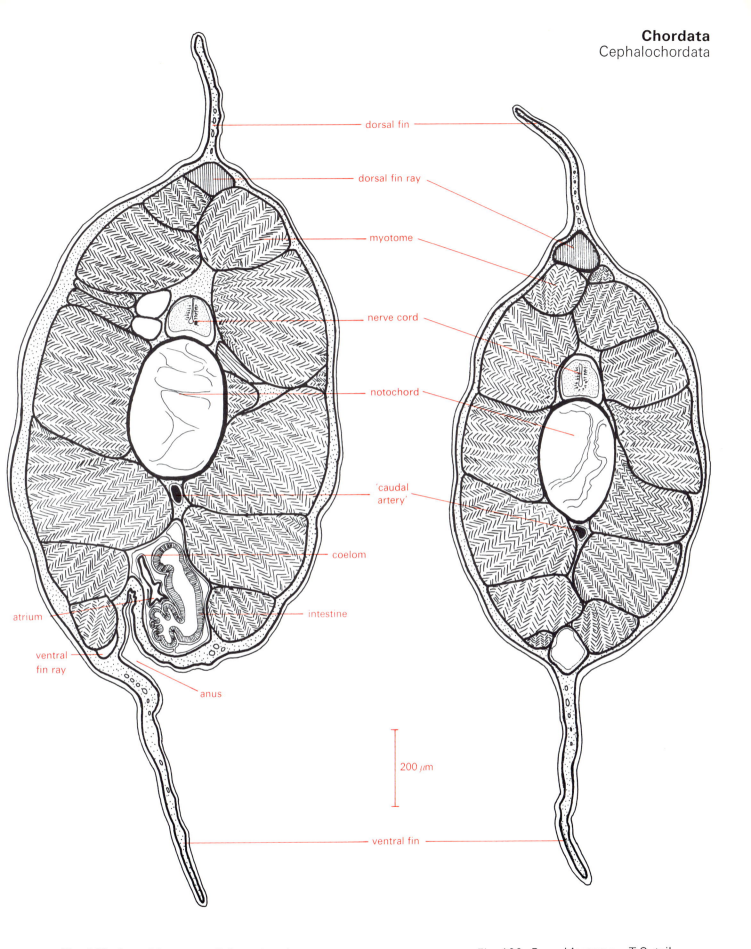

dorsal fin

dorsal fin ray

myotome

nerve cord

notochord

'caudal artery'

coelom

atrium

intestine

ventral fin ray

anus

200 μm

ventral fin

Fig. 135. *Branchiostoma* – T.S. anal region

Fig. 136. *Branchiostoma* – T.S. tail

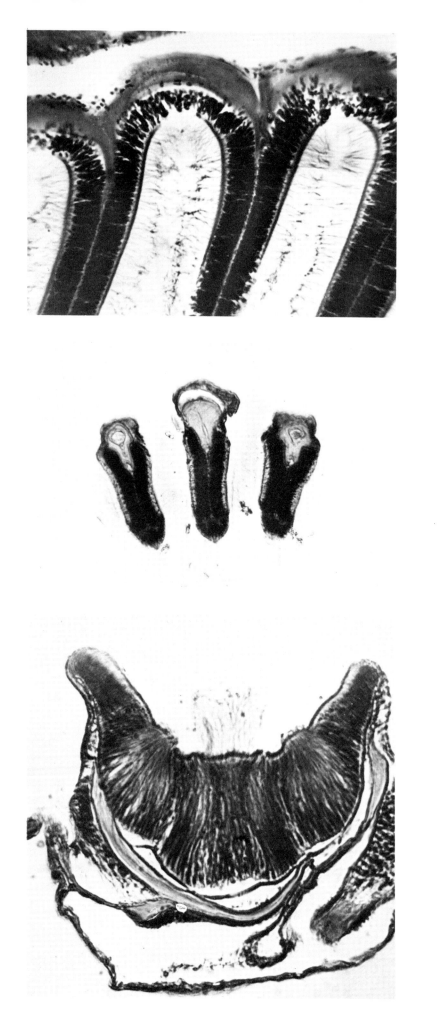
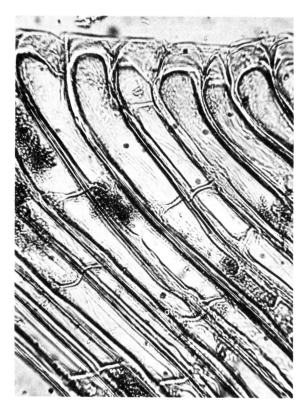
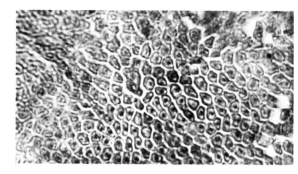
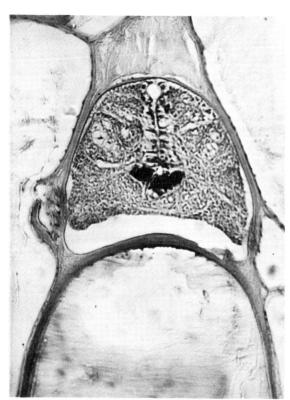

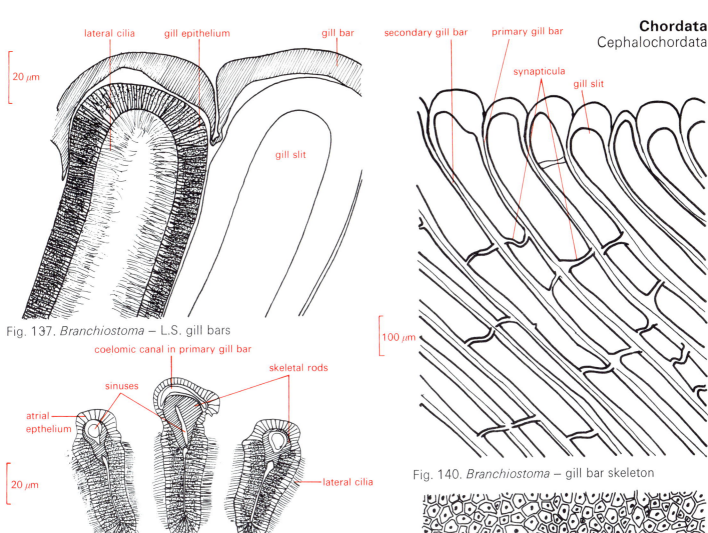

Fig. 137. *Branchiostoma* – L.S. gill bars

Fig. 138. *Branchiostoma* – V.S. gill bars

Fig. 140. *Branchiostoma* – gill bar skeleton

Fig. 141. *Branchiostoma* – surface view of skin

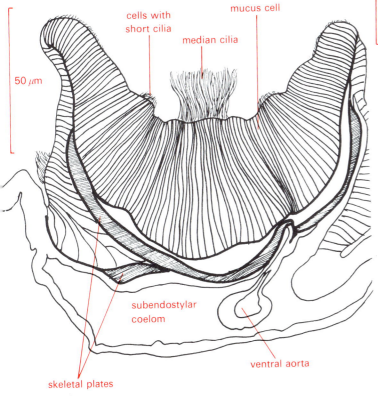

Fig. 139. *Branchiostoma* – V.S. endostyle

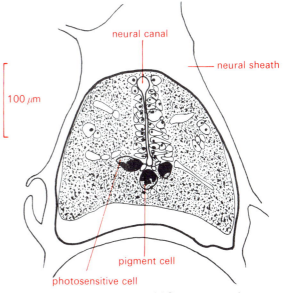

Fig. 142. *Branchiostoma* – V.S. nerve cord

Index